Tupolev Tu-Blackjack

Russia's Answer to the B-1

Yefim Gordon

Original translation by Dmitriy Komissarov

Midland Publishing

Tupolev Tu-160: Russia's Answer to the B-1
© 2003 Yefim Gordon
ISBN 1 85780 147 4

Published by Midland Publishing
4 Watling Drive, Hinckley, LE10 3EY, England
Tel: 01455 254 490   Fax: 01455 254 495
E-mail: midlandbooks@compuserve.com

Midland Publishing is an imprint of
Ian Allan Publishing Ltd

Worldwide distribution (except North America):
Midland Counties Publications
4 Watling Drive, Hinckley, LE10 3EY, England
Telephone: 01455 254 450   Fax: 01455 233 737
E-mail: midlandbooks@compuserve.com
www.midlandcountiessuperstore.com

North American trade distribution:
Specialty Press Publishers & Wholesalers Inc.
39966 Grand Avenue, North Branch, MN 55056, USA
Tel: 651 277 1400   Fax: 651 277 1203
Toll free telephone: 800 895 4585
www.specialtypress.com

© 2003 Midland Publishing
Design concept and layout
by Polygon Press Ltd. (Moscow, Russia)
Colour artwork by Vasiliy V. Zolotov
Line drawings: Yefim Gordon archive

This book is illustrated with photos by Yefim Gordon, Sergey Skrynnikov, Sergey Popsuyevich, Victor Drushlyakov, Hugo Mambour, Tupolev Joint-Stock Company, the archive of Il'dar Bedretdinov, *Ves'nik Aviatsii i Kosmonavtiki* (Aerospace Herald), *Nezavisimoye Voyennoye Obozreniye* (Independent Military Review), *World Air Power Journal*, the Russian Aviation Research Trust and the ITAR-TASS News Agency

Printed in England by Ian Allan Printing Ltd
Riverdene Business Park, Molesey Road,
Hersham, Surrey, KT12 4RG

All rights reserved. No part of this publication may be reproduced, stored in a retrieval system, transmitted in any form or by any means, electronic, mechanical or photo-copied, recorded or otherwise, without the written permission of the publishers.

Contents

Introduction 3
1. The Great Contest –
   You Win Only to Lose 5
2. Taking Shape –
   From M-18 to Tu-160 27
3. Tests and Production –
   The *Blackjack* Becomes Reality 35
4. The Tu-160 in Detail 57
5. In Soviet Service –
   And in Later Days 67
6. *Blackjack* vs 'Bone' –
   Equals or Not? 113
   Line drawings 118
   Colour drawings 122

Title page: The Tu-160 makes a distinctive impression when approaching head-on with the wings at full sweep.

Below: A nice study of '12 Red', the first Tu-160 to be demonstrated to the West. Note the five auxiliary blow-in doors; a sixth door was added on late aircraft.

Introduction

It is a generally known fact that the Soviet Union and the United States were allies during the Second World War. Soon after the end of the hostilities, however, Europe found itself divided according to the strategic interests of the two superpowers and co-operation gave way to animosity. The North Atlantic Treaty Organisation (NATO) was formed in 1949, followed in 1955 by its Soviet counterpart, the Warsaw Pact military bloc. The two were in constant confrontation in the more than three decades that followed; more than once it seemed that the Cold War going on between the East and the West would escalate into a full-blown 'hot' war – the Third World War. In short, the Soviet Union and the United States were, as a line from an early 1980s song went, *cowboys and Indians of today*.

Spurred by politicians and generals on both sides of the Iron Curtain, the arms race offered huge opportunities for developing new technologies but took a heavy toll on the economic health of the Soviet bloc nations which would not be outdone by the West in anything – especially in defence matters. The confrontation between the East and the West accelerated military technology development considerably, especially as far as missile and aircraft design was concerned. For many years it was a neck-and-neck race; now and then one of the superpowers would gain a lead and the opposing side would then strive to catch up. For instance, in the late 1950s and early 1960s the Soviet Union led the way in intercontinental ballistic missile (ICBM) development, whereas the USA placed its bet on strategic bombers. The military balance between the two nations and the two military blocs remained virtually until the early 1990s – that is, until the demise of the Soviet Union.

Quite often the decisions on new weapons systems development taken by the Soviet political and military elite were out of touch with reality, not correlating with the nation's economic capabilities and running contrary to common sense. On the other hand, Soviet designers were quite a match for their Western colleagues (and often bettered them), and their creative endeavours were mostly hampered by political decisions and the myopic views of the military top brass.

The field of strategic bomber design is a prime example. The design bureaux led by Andrey Nikolayevich Tupolev (OKB-156), Vladimir Mikhaïlovich Myasishchev (OKB-23), Robert Lyudvigovich Bartini (OKB-938) and Pavel Osipovich Sukhoi (OKB-51) evolved many projects which were often ahead of their time but never got off the drawing board. In the Tupolev OKB these included the recently declassified 'aircraft 125' (Tu-125) and 'aircraft 135' (Tu-135) strategic missile carriers, information on which has recently been published in the popular press. The Soviet Union was actively pursuing ICBM programmes, especially in the years when Nikita Sergeyevich Khruschchov with his famous 'missile itch' was running the country and heavy strike aircraft were out of favour. Few were destined to reach the hardware stage and even fewer were to see the end of their test programmes – not because they showed disappointing performance, mind you; on the contrary, they were much too advanced and thus were considered a threat by the ICBM lobby!

In keeping with this policy all work on the Myasishchev M-50 and M-52 strategic bombers (NATO reporting name *Bounder*) was terminated in the early 1960s and the Myasishchev OKB itself was liquidated (fortunately to be reborn later). The Sukhoi T-4

Above and below: a wooden wind tunnel model depicting one of the configurations of the projected Tu-135 missile carrier. The engines are placed in paired nacelles; note the anhedral on the outer wings.

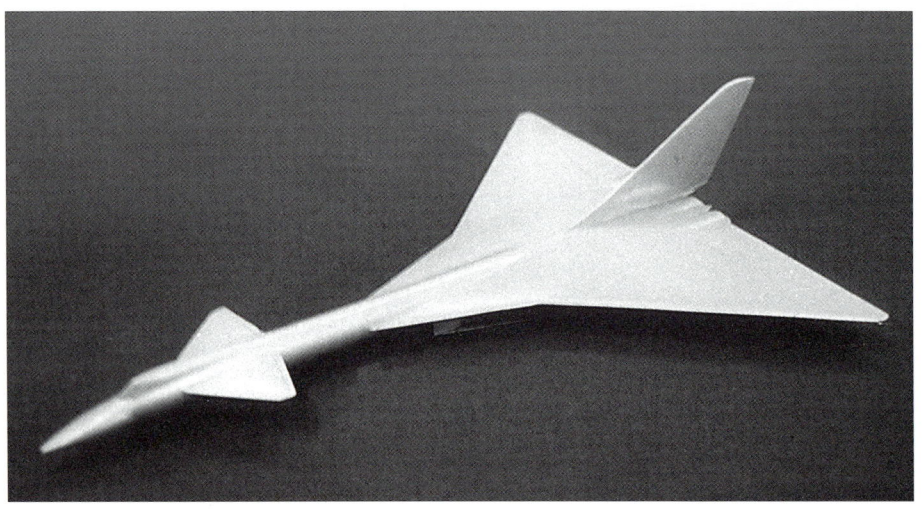

Above: Another project configuration of the Tu-135 with a single fin and rudder; the engines are located in a single package under the wing centre section.

strategic missile carrier (aka *iz**del**iye* (product) 100), with an advanced all-titanium airframe which had begun its test programme successfully was similarly victimised in 1974 after making only ten test flights. Thus, while possessing large nuclear attack assets in the form of ICBMs, by the mid-1970s the Soviet Union had only a small strategic bomber arm equipped with Tupolev Tu-95 *Bear* and Myasishchev M-4 (3M) *Bison* bombers. These obsolescent subsonic aircraft stood no chance against the well-equipped, modern air defences of the 'potential adversary' (ie, NATO). Conversely, the Americans consistently developed and refined the aviation component of their nuclear attack force.

It was not until 1967, several years after the end of the Khruschchov era, that the Soviet military turned their attention to the much-neglected strategic bomber arm of the Soviet Air Force (VVS – *Vo**yen**no-voz**doosh**nyye* *see**ly*). This change of heart was prompted by the USA's decision to launch the AMSA (Advanced Manned Strategic Aircraft) programme, which emerged as the Rockwell International B-1 bomber. A request for proposals (RFP) was issued by the Soviet government for the development of a new multi-mission, multi-mode strike aircraft possessing intercontinental range; this culminated in the now world-famous Tu-160 bomber/missile strike aircraft known to the West – and to many people in its home country – under the NATO reporting name *Blackjack*. This book tells the story of how this advanced Soviet weapons system was born amid a lot of contention and devious schemes.

Acknowledgements

The author wishes to express his gratitude to a number authors whose publications were used as sources in the preparation of this book, notably Vladimir G. Rigmant (Tupolev OKB) whose in-depth research of the subject deserves special mention. Thanks go also to Dmitriy Komissarov who, with his usual attention to detail, made useful additions to the text at the translation stage. Finally, the author would like to thank the Belgian photographer Hugo Mambour and the Ukrainian photographer Sergey Popsuyevich, both of whom supplied excellent photos used in this book.

The highly sophisticated Sukhoi T-4 (*izdeliye* 100) strategic bomber was ahead of its time and was victimised by the fight for orders within the Ministry of Aircraft Industry; however, experience accumulated with it enabled the Sukhoi OKB to develop the T-4MS bomber project. Illustrated is the sole prototype ('101 Yellow').

Chapter 1

The Great Contest

You Win Only to Lose

On 28th November 1967 the Soviet Union's Council of Ministers issued directive No. 1098-378 ordering the commencement of design work on what was referred to as a strategic intercontinental aircraft; this was the RFP mentioned earlier. The design bureaux participating in the tender were required to develop a delivery vehicle possessing outstanding performance. Suffice it to say that cruising speed at 18,000 m (≈ 59,000 ft) was specified as 3,200-3,500 km/h (1,987-2,174 mph; 1,730-1,890 kts); range in this mode was 11,000-13,000 km (6,830-8,075 miles). Maximum range in high-altitude subsonic cruise and at sea level was to be 16,000-18,000 km (9,940-11,180 miles) and 11,000-13,000 km respectively. The armament was to vary according to the nature of the mission, consisting of air-to-surface missiles – four Kh-45 *Molniya* (Lightning) missiles, or twenty-four Kh-2000 missiles etc. – or free-fall and guided bombs of various types. The missiles were products of the Moscow-based Raduga design bureau (MKB Raduga, pronounced *rahdooga* – Rainbow), although development of the Kh-45 had been initiated by the Sukhoi OKB as the main weapon for the T-4. The specified maximum ordnance load was 45 tons (99,200 lb).

Two design bureaux – the Sukhoi OKB and the Myasishchev OKB (which, as already mentioned, was reborn in the mid-1960s) – took on the task; the Tupolev OKB was probably not in a position to join the contest at this stage, having other important programmes to complete. Proceeding from the government directive and the provisional operational requirement issued by the VVS, the two OKBs had completed their advanced development projects in the early 1970s. Both contenders were four-engined aircraft with variable-geometry (VG) wings but utilised completely different aerodynamic layouts.

The Sukhoi OKB started work on a two-mode strategic bomber bearing the manufacturer's designation T-4MS or *izdeliye* 200. The engineers paid special attention to ensuring maximum commonality with the earlier T-4 *sans suffixe* (*izdeliye* 100). Among other things, the powerplant consisting of four 16,000-kgp (35,273-lb st) Kolesov RD36-41 afterburning turbojets was retained. So were the predecessor's systems and equipment, structural materials, detail design features and the manufacturing technologies mastered during the T-4 programme.

Several general arrangements of the T-4MS were studied at the preliminary design (PD) stage. At first the engineers considered simply scaling up the earlier T-4M project featuring VG wings (aka *izdeliye* 100I; the I stood for *izmenyayemaya strelovidnost'* – variable sweep). However, they soon realised it was a bad idea; this approach led to a dramatic increase in the bomber's overall dimensions and structural weight while still offering insufficient internal space for weapons stowage.

The OKB had to seek other solutions. The general arrangement of the future T-4MS had to meet the following main criteria. The internal volume had to be maximised while keeping the surface area (and hence drag) to a minimum. The weapons bays had to be capacious enough to accommodate the required range of armament. The structure had to be as stiff as possible to permit high-speed ultra-low-level operations. (This flight mode, which increased the chances of penetrating the enemy's air defences, placed high demands on structural strength because in low-level flight turbulence might occur and terrain

Two views of a display model of the Sukhoi T-4MS (*izdeliye* 200), showing the flattened triangular lifting body, the movable outer wings at minimum sweep, the engine placement and the huge nose radome.

Project specifications of the Sukhoi T-4MS bomber

Powerplant:	
project Stage A	4 x Kolesov RD36-41
project Stage B	4 x K-101
Engine power, kgp (lb st):	
project Stage A	4 x 16,000 (4 x 35,270)
project Stage B	4 x 20,000 (4 x 44,090)
Thrust/weight ratio at take-off power:	
project Stage A	0.38
project Stage B	0.47
Wing loading for overall wing area, kg/m² (lb/sq ft)	335 (68.6)
Length overall	41.2 m (135 ft 2 in)
Height on ground	8.0 m (26 ft 3 in)
Wing span:	
inner wings	14.4 m (47 ft 3 in)
at minimum sweep (30°)	40.8 m (133 ft 10½ in)
at maximum sweep (72°)	25.0 m (82 ft ¼ in)
Landing gear track	6.0 m (19 ft 8 ¼ in)
Landing gear wheelbase	12.0 m 39 ft 4 ½ in)
Outer wing area, m² (sq ft):	
at maximum sweep	73.1 (786)
at minimum sweep	97.5 (1,048)
Inner wing area, m² (sq ft)	409.2 (4,400)
Overall wing area, m² (sq ft):	
at maximum sweep	482.3 (5,186)
at minimum sweep	506.8 (5,449)
Inner wing leading edge sweep	
Outer wing leading edge sweep:	
at maximum sweep	72°
at minimum sweep	30°
Aspect ratio with respect to overall wing area:	
at maximum sweep	1.14
at minimum sweep	3.3
Empty weight, kg (lb)	123,000 (271,160)
Maximum take-off weight, kg (lb)	170,000 (374,780)
Normal take-off weight, kg (lb)	170,000 (374,780)
Internal fuel load, kg (lb)	97,000 (213,845)
Ordnance load, kg (lb):	
normal (internal)	9,000 (19,840)
maximum (internal bays and external hardpoints)*	45,000 (99,200)
Top speed, km/h (mph; kts):	
at sea level	1,100 (683; 595)
at altitude	3,200 (1,987; 1,730)
Cruising speed, km/h (mph; kts):	
above 18,000 m (59,000 ft)	3,000-3,200 (1,863-1,987; 1,621-1,730)
at medium altitude	800-900 (497-559; 432-486)
at sea level	850 (528; 459)
Maximum range with K-101 engines at cruising speed with normal warload, internal fuel only, km (miles):	
above 18,000 m (59,000 ft)	9,000 (5,590)
at medium altitude	14,000 (8,695)
Take-off run, m (ft)	1,100 (3,610)
Landing run, m (ft)	950 (3,120)
Crew	3
Armament:	
long-range air-to-surface missiles	4 x Kh-45
short-range air-to-surface missiles	24 x Kh-15
bombs/total weight, kg (lb)	45,000 (99,200)

* with partial fuel load

avoidance manoeuvres may be needed.) The powerplant had to be located externally so as to facilitate eventual re-engining (ie, buried engines were out of the question because integrating new engines might require drastic structural changes). Finally, the layout had to offer the potential of continuously improving the aircraft's performance and handling.

As work progressed on the final versions of the T-4M project utilising the so-called integral or blended wing/body (BWB) layout where the fuselage contributes a large amount of lift, Sukhoi OKB engineers decided that a 'flying wing' BWB layout would meet the demands described above. A while earlier, their colleagues at the Tupolev OKB had arrived at the same conclusion. Unlike Tupolev, however, the Sukhoi OKB proposed variable-geometry wings with movable outer portions of relatively small area. This 'flying wing'/'swing wing' combination was probably unique in aircraft design practice.

Known in-house as 'version 2B', the 'flying wing'/'swing wing' layout was developed in August 1970 by engineer L. I. Bondarenko. In due course it was approved by PD section chief Oleg S. Samoylovich, then by the T-4MS's chief project engineer N. S. Chernikov and finally by General Designer Pavel O. Sukhoi, and served as the basis for the advanced development project.

Wind tunnel tests at the Central Aerodynamics & Hydrodynamics Institute named after Nikolay Ye. Zhukovskiy (TsAGI – *Tsentrahl'nyy aero- i ghidrodinamicheskiy institoot*) showed that the chosen layout offered a high lift/drag ratio in both subsonic and supersonic modes. Actually 'high' is too modest a description; the results were truly fantastic – an L/D ratio of 17.5 at Mach 0.8 and 7.3 at Mach 3.0. The new integral layout also took care of aeroelasticity problems. The limited area of the movable outer wings, coupled with the stiff structure of the wing centre section/fuselage (lifting body), enabled high-speed flight at low altitude.

Work on defining and refining the advanced development project (ADP) of *izdeliye* 200 to the degree when it could be submitted for the tender continued throughout 1971. Wind tunnel models were manufactured that year, allowing different versions of the lifting body, outer wings, vertical and horizontal tail to be tested in TsAGI's wind tunnels. The tests showed that the T-4MS was catastrophically unstable because the centre of gravity shifted too radically when wing sweep was altered. Chief project engineer N. S. Chernikov decided to alter the layout. As a result, several project versions emerged featuring an extended nose and additional (conventionally placed) horizontal tail surfaces; one of them (version 8) had a needle-sharp nosecone.

The configuration selected eventually featured an extended forward fuselage with an extremely streamlined flight deck canopy so that the upper fuselage contour was virtually unbroken; apart from this, there were no changes as compared to the original ADP. The T-4MS project was completed in September 1971. The bomber's design specifications are given in the table on page 6.

Now we will turn our attention to the competitor. The Myasishchev OKB (officially known as EMZ – *Eksperimentahl'nyy mashi-nostroitel'nyy zavod*, experimental machinery plant) received orders from the Ministry of Aircraft Industry (MAP – *Ministerstvo aviatsionnoy promyshlennosti*) to develop a PD project of a strategic multi-mode missile carrier as far back as 1968. This was to be a multi-mission strike aircraft with three distinct operational configurations.

The EMZ design team set to work with a will, ignited by the enthusiastic approach of its leader, Vladimir M. Myasishchev. The project was known in-house as **tema dvahdsat'** ('subject 20'), alias the M-20 multi-mode bomber/missile carrier. The basic strike/reconnaissance configuration was intended for attacking remote targets of strategic importance with nuclear-tipped missiles or bombs and performing strategic reconnaissance. The second configuration was a counter-air version designed to seek and destroy transport aircraft flying transoceanic routes and airborne early warning (AEW) aircraft. Finally, the third version was a long-range anti-submarine warfare (ASW) aircraft intended to seek and destroy submarines at up to 5,000-5,500 km (2,700-2,970 nm) away from base. The aircraft's maximum range in subsonic cruise was specified as 16,000-18,000 km (9,940-11,180 miles).

As he did before, V. M. Myasishchev believed the creation of a heavy fast strike aircraft to be the main task of his reborn OKB. After the PD work on 'subject 20' had been completed he succeeded in getting the EMZ entered into the tender for the supersonic strategic missile carrier. MAP orders to this effect were issued on 15th September 1969 (No. 285), 17th September 1970 (No. 134) and 9th October 1970 (No. 321). The OKB started work on a new project – **tema vosemnahdsat'** ('subject 18'), alias M-18.

On 15th February 1971 Myasishchev delivered a report to the assembled representatives of various research establishments and OKBs, describing the progress the EMZ had made on the programme jointly with TsAGI and several research institutes within the frameworks of the Ministry of Defence, the Ministry of Electronics Industry and the Ministry of Defence Industry. In his report he pointed out that the general operational requirement (GOR) for the new

Above and below: With the wings fully swept back, the T-4MS had an almost perfectly triangular shape. The entire trailing edge of the lifting body between the engine nozzles was a four-section elevator.

An artist's impression of the T-4MS (*izdeliye* 200) in flight. Note the small area of the flight deck glazing.

Above: Another desktop model of the T-4MS. This one features a larger flight deck glazing area.

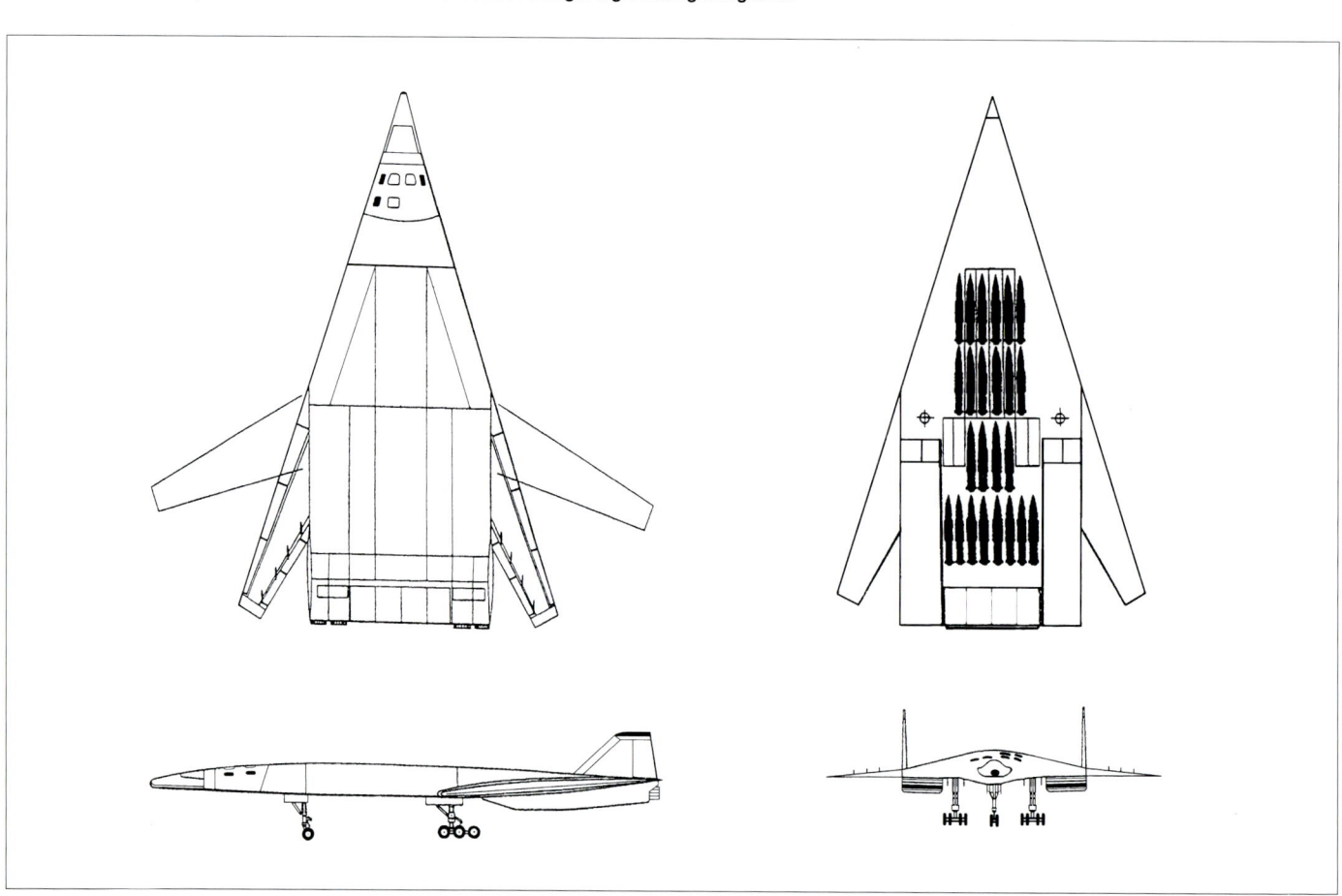

A three-view of the T-4MS illustrating the stalky undercarriage with 12-wheel main bogies. The diagram on the right shows the placement of the 24 Raduga Kh-15 missiles.

Above: A display model of the Myasishchev M-20 in one of its many project configurations featuring a tail-first layout. Note the three eight-wheel main gear bogies and the bifurcated air intake serving all four engines.

This version of the M-20 featured a common nacelle for all four engines; the bifurcated air intake with vertical flow control ramps divided into channels flanking the weapons bay.

Above: The same M-20 model with the wings at maximum sweep.
Below: The tailcone housed a three-cannon barbette and a gun ranging radar. The vertical tail was cut away at the base to give an adequate field of fire.

Above: With the wings fully swept back the M-20 had a distinctive arrow-like shape. Note the high-set canards.
Below: Even in the fully forward position, the wing sweep angle was fairly high.

12

Previous page, top: A desktop model showing one of the project configurations of the Myasishchev M-18 (probably the definitive one). Apart from the design of the tail unit, the aircraft bore a striking resemblance to the future Tu-160, with engines in paired nacelles to leave the centreline free for weapons bays. Here the wings are shown at maximum sweep.

Previous page, bottom: As distinct from the conventionally built M-20 which had a circular-section fuselage sitting on top of the wing centre section, the sleek M-18 featured a blended wing/body layout. Unlike the future Tu-160, it utilised a conventional wing glove design for swing-wing aircraft, with the outer wing trailing edges sliding inside the inner wing 'pockets'.

This page, right: In fully forward position the M-18's wings are almost unswept (another difference from the M-20), making a sharp contrast with the tail surfaces.

Below: This view illustrates the sharply swept tail surfaces with low-set stabilisers and an unbroken fin leading edge – something that would change when the project was reworked to become the Tu-160. Like the M-20, the M-18 featured defensive armament.

OKB-51 General Designer Pavel Osipovich Sukhoi. Though his design bureau mainly specialised in fighter design, Sukhoi also supervised the development of the highly advanced T-4 (*izdeliye* 100) bomber and later the T-4MS (*izdeliye* 200) which was regarded as the most promising of the projects submitted for the tender.

General Designer Vladimir Mikhaïlovich Myasishchev, head of OKB-23, pictured here in military uniform (he had the rank of Major General). His OKB specialised in bomber design, though few of his bombers were destined to see production and service; nevertheless, it was the M-18 project that ultimately served as the basis for the Tu-160.

bomber specified an increase in warload over aircraft then in service with the Soviet Air Force by a factor of 1.8, which led to a higher all-up weight. The GOR also demanded the provision of special equipment facilitating the penetration of air defences (obviously electronic countermeasures (ECM) equipment which disrupts the operation of enemy AD radars – *Auth*.), an improvement in thrust/weight ratio over existing aircraft by a factor of at least 1.5-1.7 (due to the need to be capable of operating from Class 1 unpaved airstrips) and a cruising speed of 3,000-3,200 km/h (1,863-1,987 mph; 1,620-1,730 kts).

All of this, according to V. M. Myasishchev's own calculations and those of his staff, reduced the aircraft's range by 28-30%.

The OKB undertook a lot of research and development work under the strategic multi-mission strike aircraft programme; much of it was performed jointly with TsAGI. This included up to 1,200 hours of computer analysis of different airframe layouts; much effort was spent on flight dynamics and controllability studies in various flight modes. The size and weight of the aircraft were optimised, proceeding from different all-up weights (ranging from 150 to 300 tons/330,690 to 661,375 lb) and dimensions. Heat transfer and heat rejection quotients were studied in TsAGI's T-33 wind tunnel, using models.

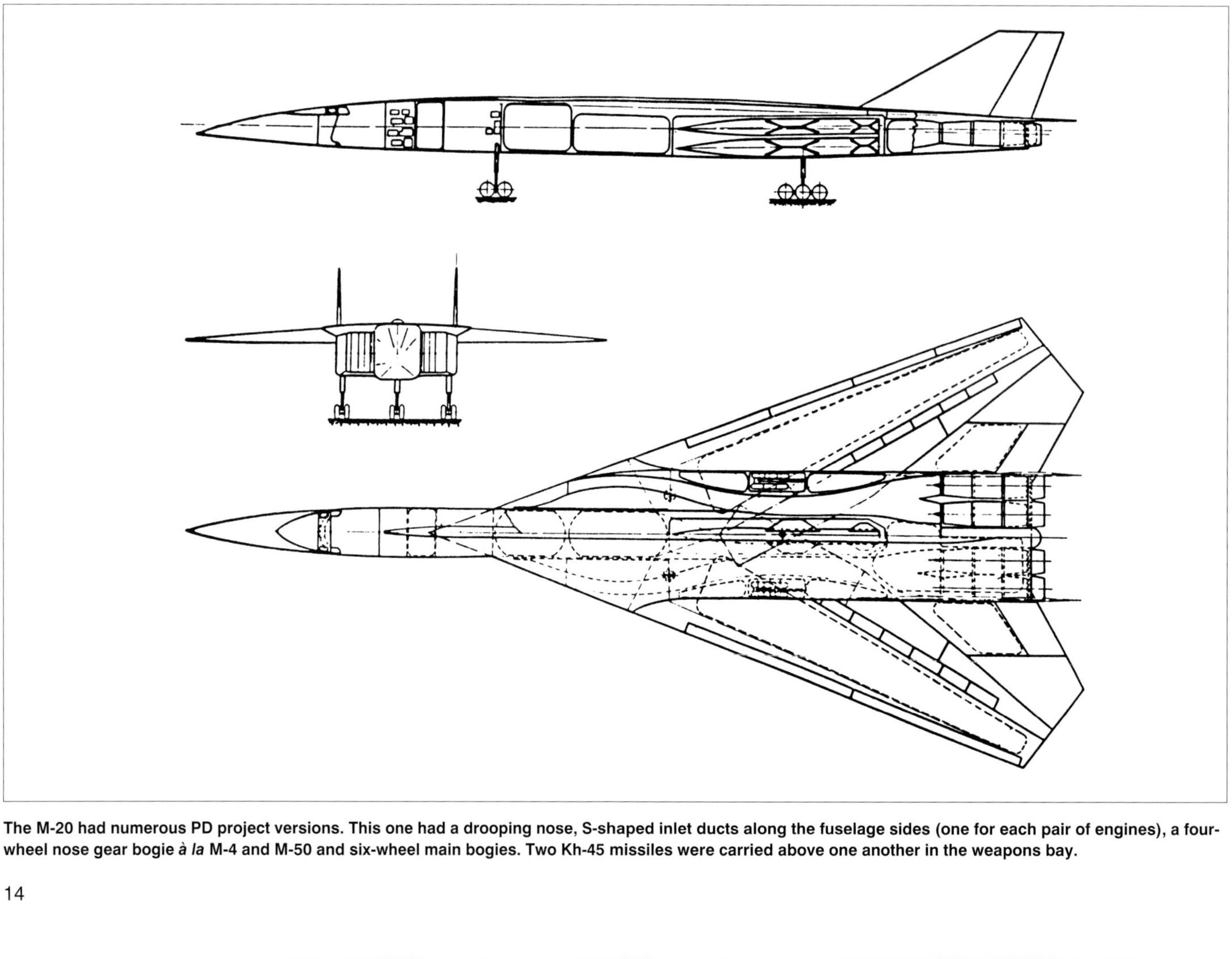

The M-20 had numerous PD project versions. This one had a drooping nose, S-shaped inlet ducts along the fuselage sides (one for each pair of engines), a four-wheel nose gear bogie *à la* M-4 and M-50 and six-wheel main bogies. Two Kh-45 missiles were carried above one another in the weapons bay.

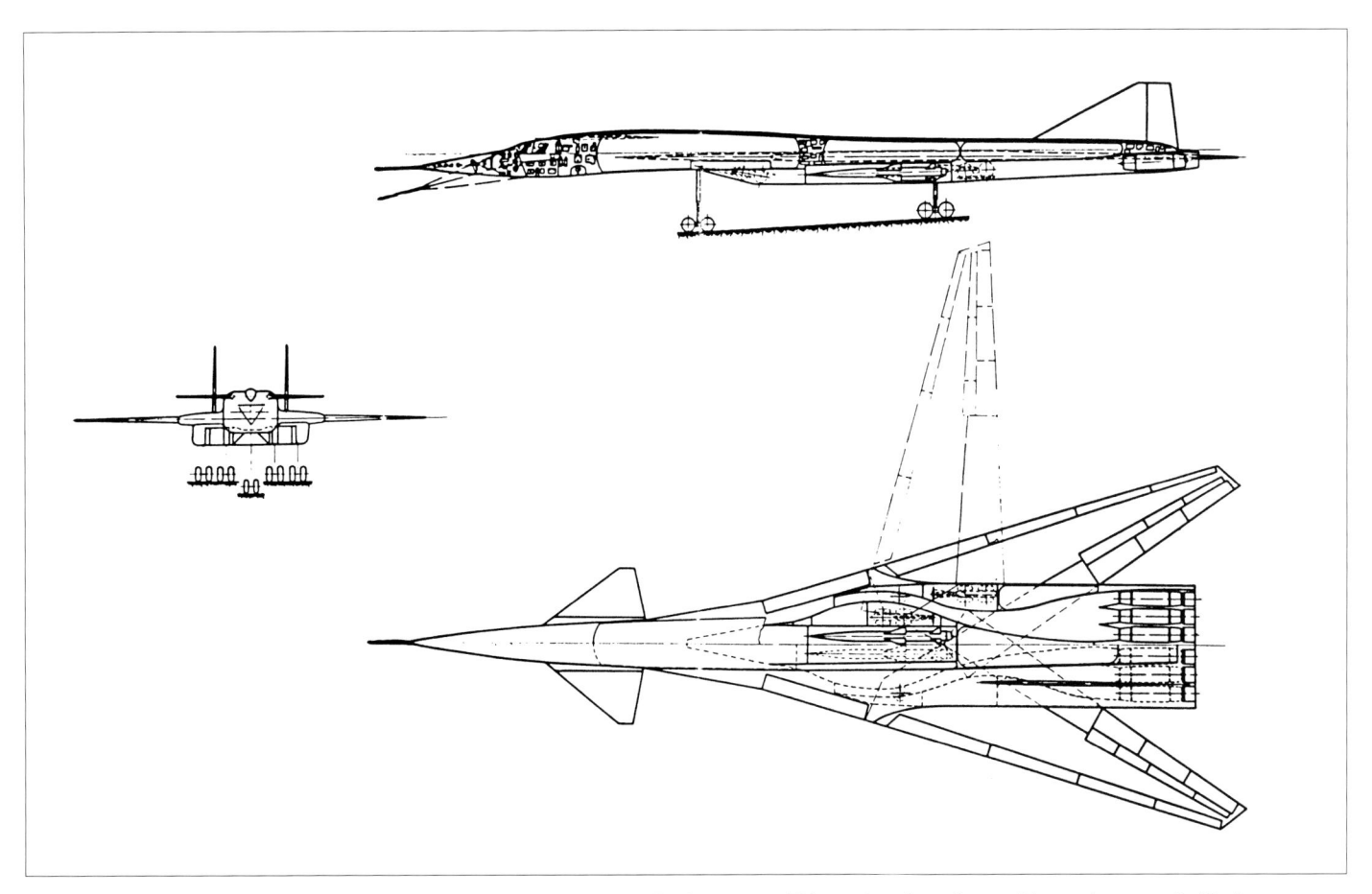

Above: Another version of the M-20, again with a drooping nose and a four-wheel nose gear. This one has six engines and four main gear units (the inner ones retract forward and the outer ones aft); the two missiles are carried side by side. Note that this version features a BWB layout and all-movable canards.

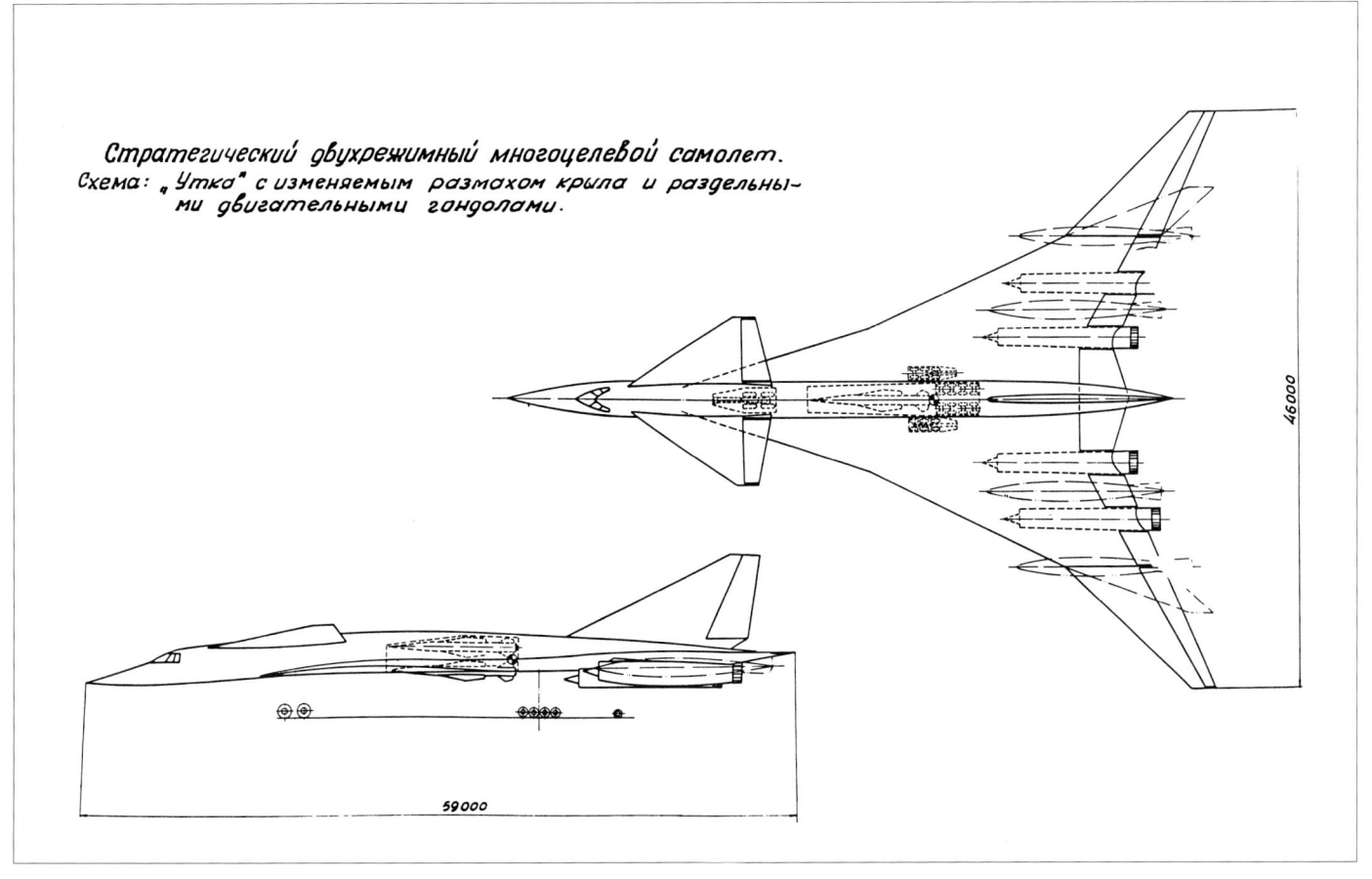

Стратегический двухрежимный многоцелевой самолет.
Схема: „Утка" с изменяемым размахом крыла и раздельными двигательными гондолами.

This version of the M-20 featured drooping wingtips and a bicycle landing gear with outrigger struts retracting into pods at mid-span where the wing hinges are. The main gear had four-axle units retracting aft and three-axle units retracting forward into the wing roots. Drop tanks were carried between the engine nacelles.

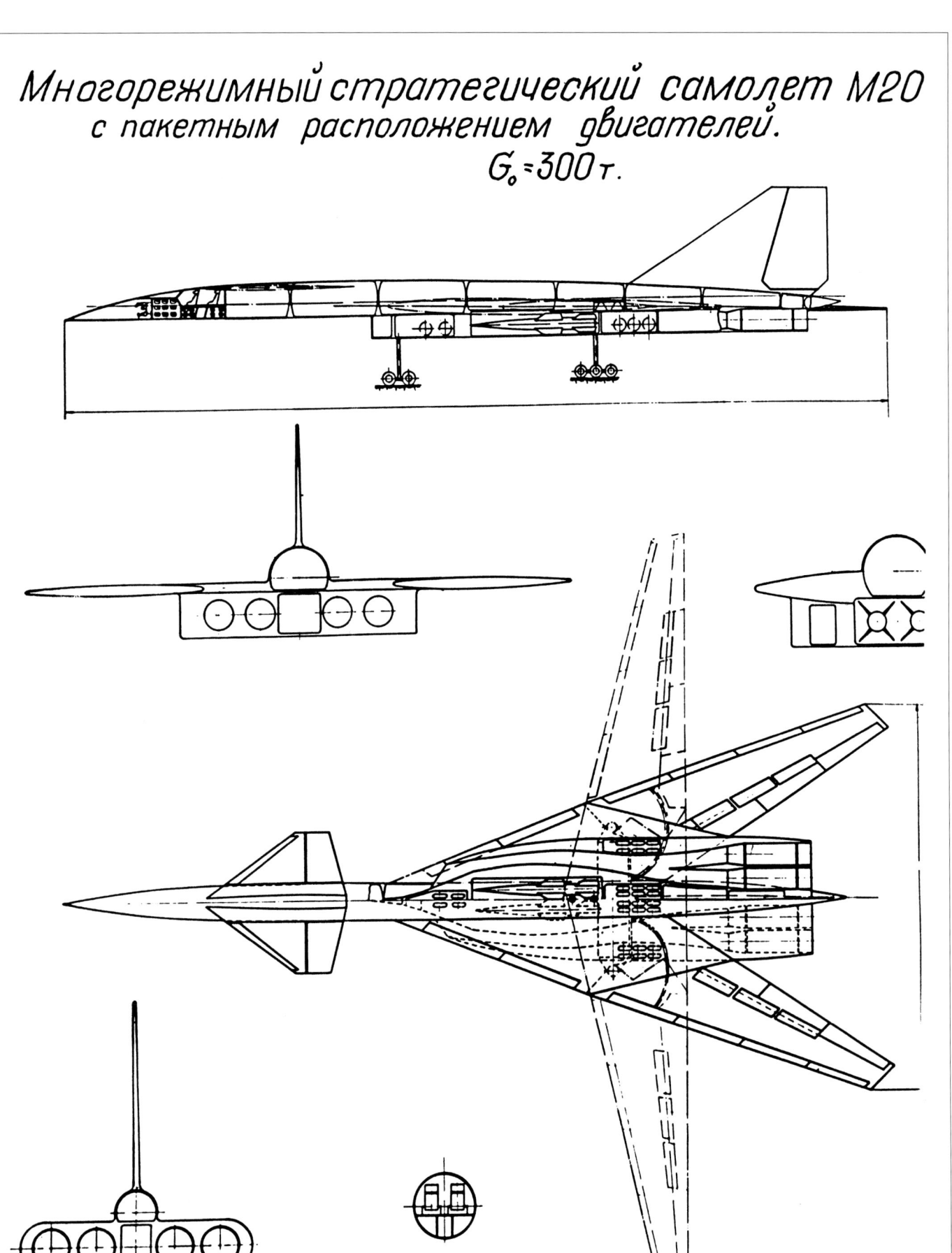

Yet another version of the M-20 with a 300-ton (661,375-lb) take-off weight; it is remarkably similar to the one on pages 9-11, except for the landing gear design. Like all the other drawings on pages 14-19, this is an original Myasishchev OKB drawing.

Многорежимный стратегический самолет М20 с УЛО.
$G_o = 300$ т.

This version of the M-20 had four engines arranged in two vertical pairs. The main gear units retracted into the five-spar wings, the thin airfoil necessitating 12-wheel bogies. This version was to feature a boundary layer control system ensuring a laminar airflow.

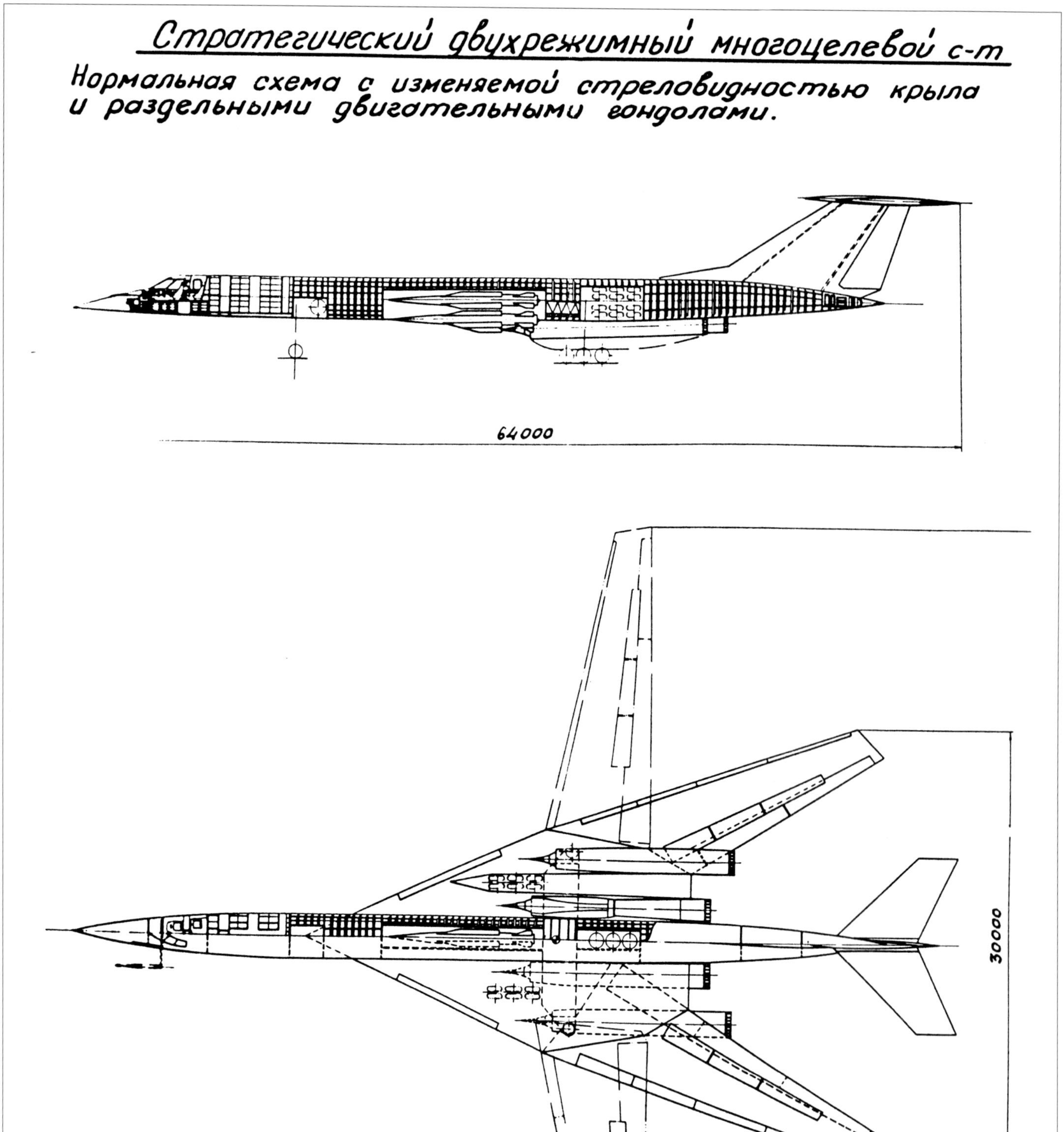

Стратегический двухрежимный многоцелевой с-т
Нормальная схема с изменяемой стреловидностью крыла и раздельными двигательными гондолами.
64000
30000

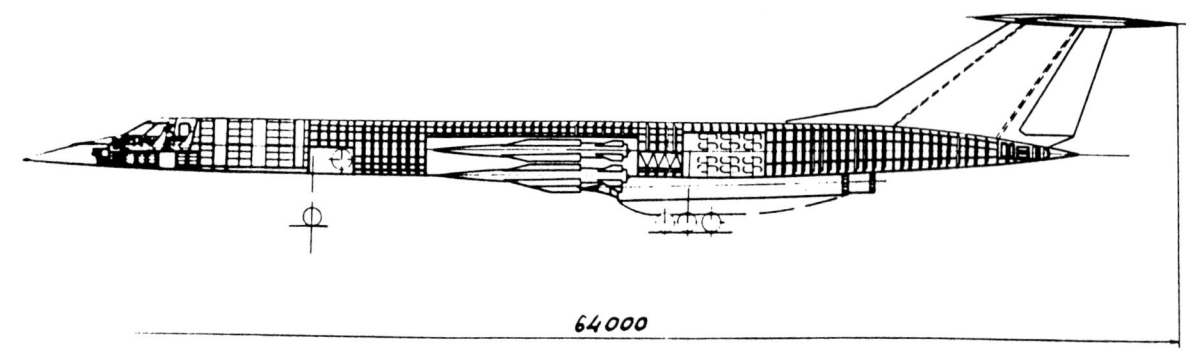

This late version of the M-20 was a stepping stone towards the M-18. It has a T-tail and four individual engine nacelles featuring circular intakes with shock cones. Two missiles are carried above each other (the lower one is semi-recessed); two main gear units retract into the wings and two more into the fuselage.

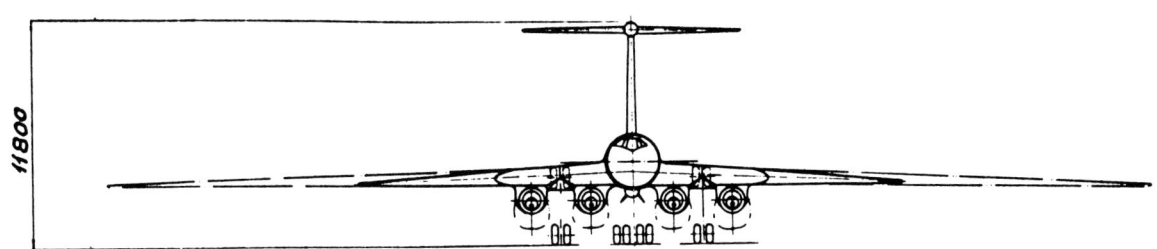

Основные данные

		Прототип 1975 г
1. Вес с-та максимальный т		300
2. Относительный вес топлива %		57,9
3. Удельная нагрузка на крыло кг/м² при $X_{пк} = 13°$		608
4. Тяговооруженность взлетная		0,3
5. Дальность полета максимальная км при $M = 0,8$; $H = 8-13$.		14200
6. Вес боевой нагрузки нормальной — максимальн. т		8,5 - 40
7. Длина разбега по бетону — грунту м		1600 - 3000
8. Двигатели: тип, число X, тяга в кг		ТРДДФ 4 × 22000 ген. конст. Кузнецов Н.Д.
9. Число членов экипажа		3 - 4

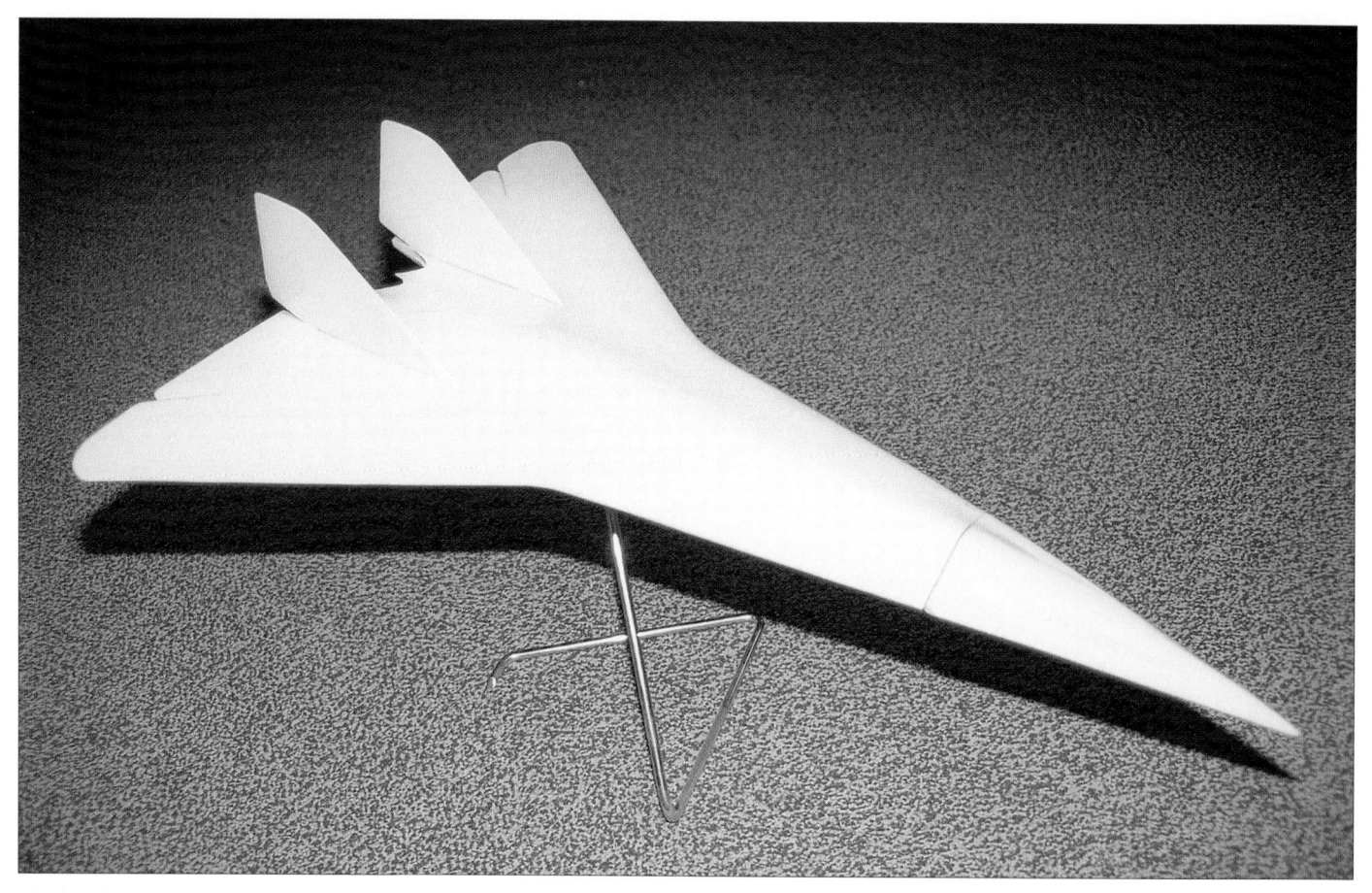

Above and opposite page: One of the first project versions of Tupolev's multi-mode strategic bomber ('aircraft 160M') featuring twin tails and cranked-delta wings. The engines are located in a single package under the wing centre section, as on the Tu-144 prototype of 1968.

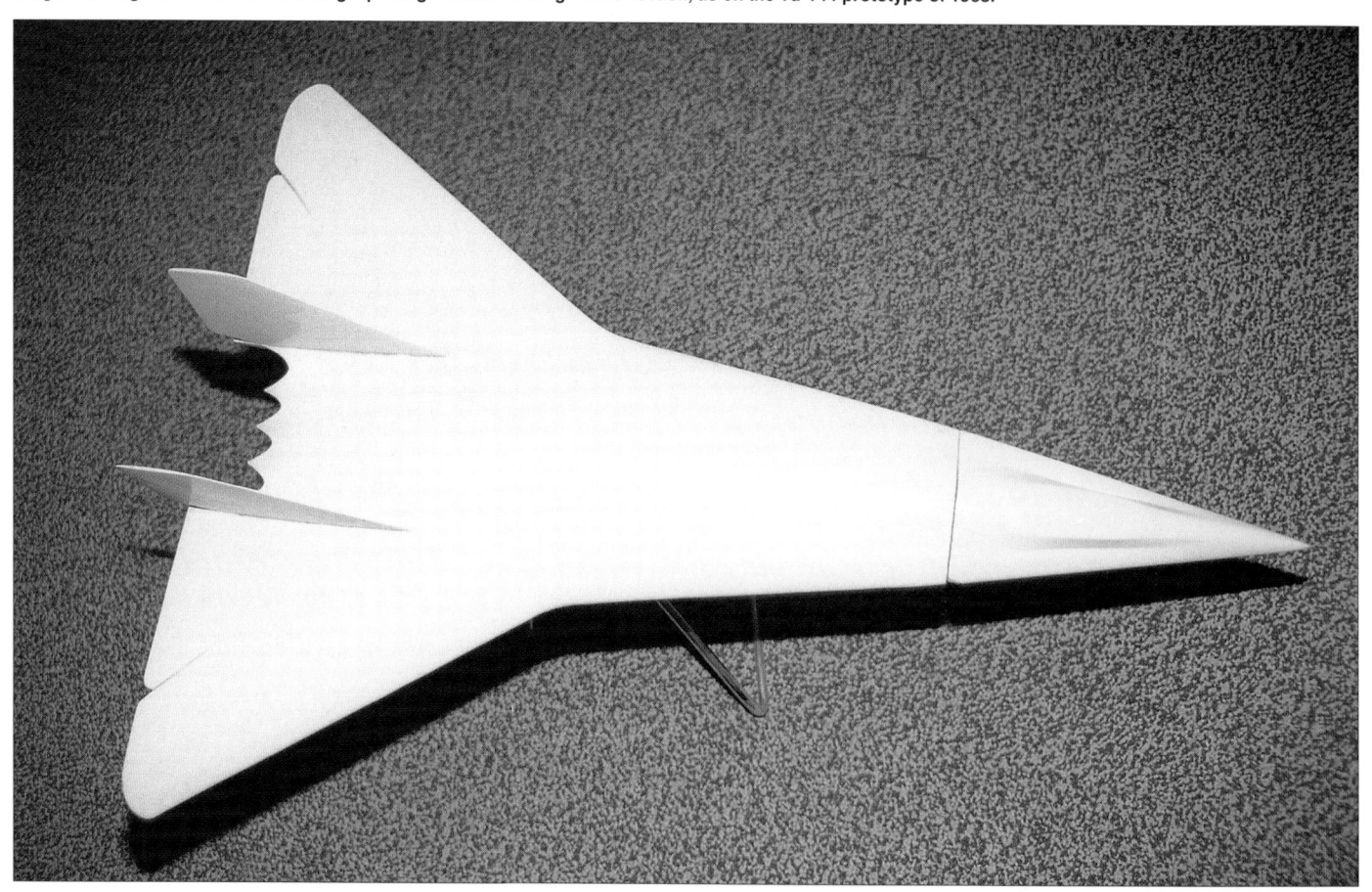

This view shows the nozzles of the four engines between the vertical tails, with 'pen nib' fairings in between.

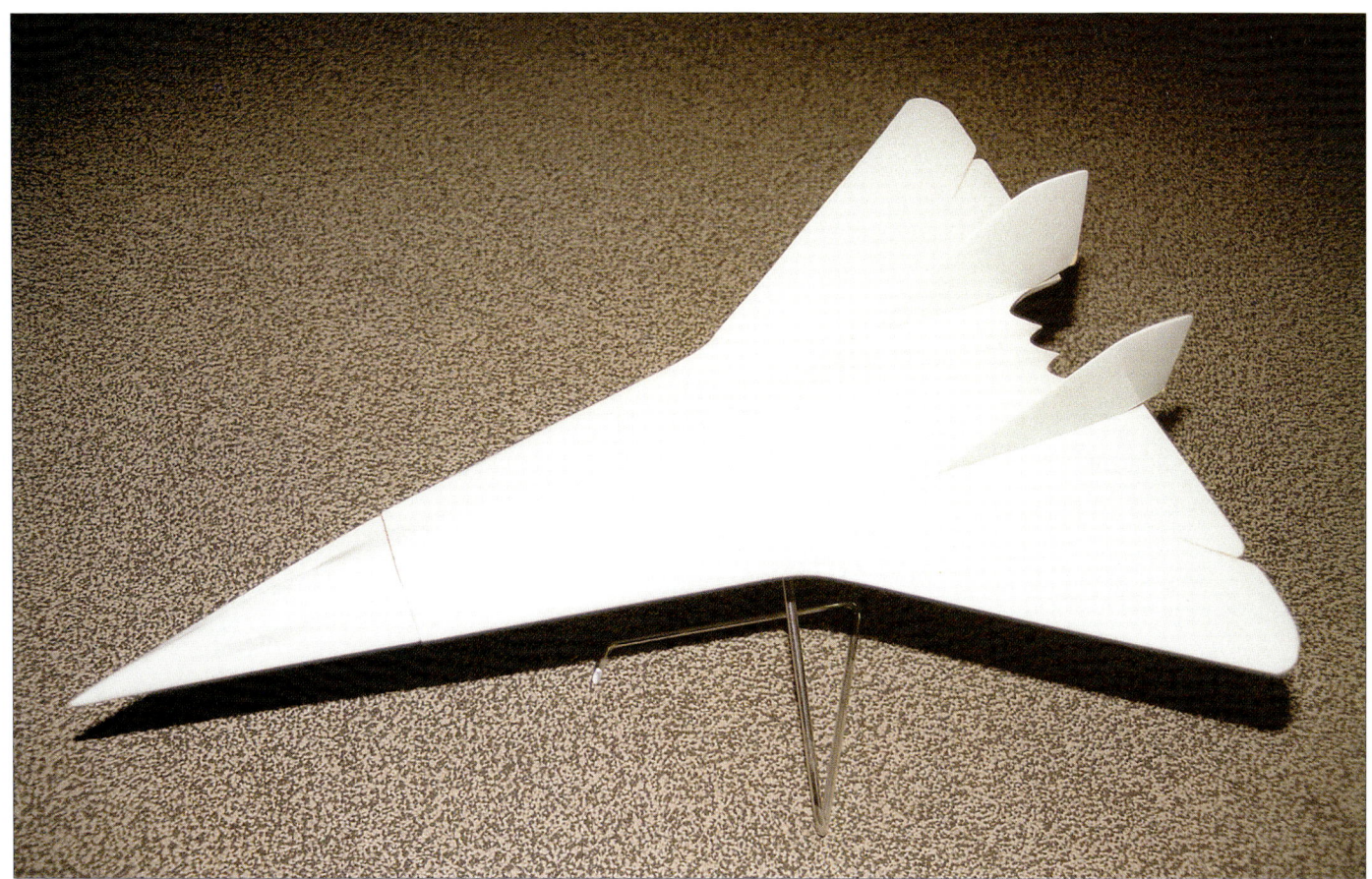

The airframe's structural strength and stiffness characteristics were determined for different aerodynamic layouts and structural materials; among other things, this required the use of the wind tunnels at the Siberian Aviation Research Institute (SibNIA – *Sibeerskiy naoochno-issledovatel'skiy institoot aviahtsii*) in Novosibirsk, in addition to TsAGI's T-203 wind tunnel. The layouts and weights of the aircraft's principal systems (flight control system, avionics, landing gear, armament, engines etc.) were studied and the best ones selected. Detail design of the wings, fuselage, landing gear and powerplant got under way at the same time.

The Myasishchev OKB (EMZ) studied several general arrangements concurrently while working on the M-18 and M-20 projects. The engineers started with the conventional layout, then tried the canard (tail-first) layout. The following options were considered, among other things:
- conventional layout with variable-geometry wings and twin tails or a single vertical tail;
- conventional layout with variable-geometry wings and a T-tail;
- tail-first layout with delta wings and delta foreplanes;
- tail-first layout with variable-geometry wings;
- tail-first layout featuring compound leading-edge sweep and drooping outer wings (as used on the North American XB-70 Valkyrie bomber);
- tailless delta layout.

The result of this research was a firm conviction that the future multi-mission strategic strike aircraft should have VG wings. The main difference between the two Myasishchev projects was that in the M-20 project the tail-first layout prevailed, whereas most versions of the M-18 project utilised a conventional layout.

General Designer Vladimir M. Myasishchev directly supervised the development of the multi-mission strike aircraft. Many other prominent designers of the recreated OKB were also heavily involved in the M-18/M-20 programme; these included Deputy General Designer G. I. Arkhangel'skiy, acting Deputy General Designer M. V. Goosarov, acting Deputy General Designer V. A. Fedotov, aerodynamics section chief A. D. Tokhoonts and many others. K. P. Lyutikov was appointed chief project engineer. Responsibilities were distributed as follows: Tokhoonts was responsible for the general arrangement, internal layout, aerodynamics and powerplant; Fedotov supervised all structural strength work, introduction of new materials and detail design. One more man, N. M. Glovatskiy, was in charge of prototype manufacturing, at the same time acting as chief engineer of the OKB's impressive experimental production facility.

The general arrangements were selected with two options in mind: an aircraft with a take-off weight around 150 tons (330,690 lb) equipped with an in-flight refuelling (IFR) system or a much larger aircraft with a TOW of 300-325 tons (661,375-716,490 lb) and no provisions for IFR. The powerplant was selected accordingly; a 150-ton aircraft would be powered by four engines delivering 12,000 kgp (26,455 lb st) each, whereas the heavy version required engines in the 22,000-25,000 kgp (48,500-55,115 lb st) thrust class. Nikolay Dmitriyevich Kuznetsov's OKB-276 based in Kuibyshev (now called *SNTK imeni Kuznetsova* – the Samara Scientific & Technical Complex named after Kuznetsov) was chosen as the engine supplier.

The crew consisted of three or four persons. The wing area ranged from 670 to 970 m² (7,204 to 10,430 sq ft), depending on the take-off weight. The principal offensive armament was to consist of two heavy air-to-surface missiles. There was no defensive armament.

In general arrangement and specific design features the M-18 closely paralleled the Rockwell International B-1 bomber. Hence it was considered to be – and promoted as – the more promising of the two Myasishchev projects (perhaps the truth is that it was simply a safer option, being less innovative than the M-20). Way ahead of all other aspects of detail design, work started on the most crucial element of the airframe – the wing pivots; the novel design of these units underwent structural and dynamic testing on

Above and below: Another early project configuration of Tupolev's strategic bomber. The similarity to the definitive version of the Tu-144 SST is obvious, but the droopable wingtips were something the Tu-144 did not have! The camouflage paint job was a trick intended to make the project more appealing to the military.

Above and below: Another model of the projected 'aircraft 160M'; this one is even more similar to the Tu-144, which is the reason why this project was rejected immediately. Note the nozzles of the Kolesov RD36-51A engines with characteristic internal cones and the tail barbette and associated gun ranging radar.

Aleksey Andreyevich Tupolev, the Tu-160's programme chief who later became General Designer of OKB-156. Several projects entered into the multi-mode strategic bomber contest were developed under his direction, but eventually the Tupolev OKB ended up reworking a project developed by a rival bureau which was declared the winner!

a scale model at TsAGI. Nine test rigs and two flying testbeds were involved in the development of the M-18. The result of this massive test and research effort was a 10% reduction in the aircraft's weight.

As already mentioned, the multi-mission, multi-mode strike aircraft projects developed by the Sukhoi and Myasishchev bureaux envisaged the strategic bomber/missile carrier role as the main one, with provision for later adaptation for the high-altitude reconnaissance or ASW roles.

After the Air Force had formulated a GOR for an advanced multi-mode strategic strike aircraft in 1969, the Soviet government decided that multiple design bureaux should compete for the order and a deadline for project submission should be set for all participants for the tender. This time the Tupolev OKB was 'invited' along with Sukhoi and Myasishchev, since Tupolev was the Soviet Union's top authority in heavy bombers and possessed the greatest expertise in this field.

Until 1970 the top executives of the Tupolev OKB attended all government meetings having to do with the new bomber strictly as observers. (Of course, one may justly say that these observers were not wholly impartial; they listened carefully and drew a few conclusions to be used later in their own bomber design efforts!) The OKB was heavily burdened with current programmes, civil as well as military (the Tu-154 *Careless* trijet medium-haul airliner, the Tu-144 *Charger* supersonic transport (SST), the Tu-22M *Backfire* bomber/missile carrier and the Tu-142 *Bear-F* long-range ASW aircraft all entered flight test in the mid-/late 1960s). Hence the leaders of the OKB did not want to bite off more than they could chew, even though the advanced multi-mode strategic strike aircraft programme clearly fitted well into the company's traditional line of work.

However, after carefully analysing the actual state and prospects of the programme, assessing its own capabilities and those of its competitors OKB-156 did indeed embark on a new strategic bomber programme, using the preliminary specifications of 1967 as a starting point. The work was performed by the design bureau's Section K under the overall co-ordination of Aleksey Andreyevich Tupolev, the General Designer's son (and future successor). Later, overall project responsibility passed to Valeriy Ivanovich Bliznyuk, a designer who had earlier participated in the development of the unbuilt 'aircraft 135' (Tu-135) strategic bomber/missile carrier, the OKB's first supersonic unmanned aerial vehicles – the '121', or Tu-121, and '123', or Tu-123 *Yastreb* (Hawk) – and the Tu-144 SST. A major contribution to the early project studies was made by Aleksandr Leonidovich Pookhov who now supervises the Tu-144LL SST technology research aircraft programme and the creation of Russia's second-generation SST, the Tu-244.

Initially known as 'aircraft 156', the new bomber was soon redesignated '160' (Tu-160). Some documents referred to it as *izdeliye* K (presumably thus named after the OKB section developing the aircraft); eventually, however, the Tu-160 came to be known as *izdeliye* 70 and this cipher is still in use today. As for the '156' (Tu-156) designation, it was later reused for a projected derivative of the Tu-154 using cryogenic fuel (liquid natural gas). Generally speaking, reusing project designations was common practice at the Tupolev OKB.

Initially the Tupolev OKB developed the *izdeliye* K unbidden, and information about it was distributed on a 'need to know' basis – only a very small group of people working at the OKB itself and the Ministry of Aircraft Industry knew about the project. Therefore the designers were allowed to do pretty much what they pleased as far as the choice of aerodynamic layout and specific design features were concerned. They decided not to reinvent the wheel, relying heavily on the unique engineering experience gained when designing the Tu-144. This served as the basis for the initial project version of Tupolev's multi-mode strike aircraft – an aircraft which differed markedly in its design from both Sukhoi's T-4MS and Myasishchev's M-18/M-20.

The extremely high demands posed by the Council of Ministers directive of 1967 meant that the engineers were facing a Mission Impossible. Therefore initially the Tupolev OKB engineers decided to select the specified maximum speed and maximum-range cruising speed as the main performance targets and take it from there. It should be noted that, concurrently with the bomber project, Section K conducted research on ways of further developing supersonic transports; this research later laid the foundation of the SPS-2 programme (*sverkhzvookovoy passazheerskiy samolyot*, SST) – the future Tu-244. It is quite logical therefore that part of this research should go into the Tu-160 programme – specifically, the general arrangement. Hence initially the bomber shared the tailless delta layout used by the SPS-1 (Tu-144) and SPS-2 (Tu-244).

Research and development data accumulated under the next-generation SST programme showed that in theory it was possible to obtain a lift/drag ratio of 7-9 in supersonic cruise and as much as 15 in subsonic cruise. This, coupled with advanced fuel-efficient engines, made reaching the specified range target – or at least approaching it – a realistic prospect. At any rate, some of the Tu-244 project documents (dated 1973) show that with turbojet engines having a specific fuel consumption of 1.23 kg/kgp·hr (lb/lb st·hr) in supersonic cruise mode the airliner was expected to have a maximum range of 8,000 km (4,970 miles). Given adequate engine power, the tailless delta layout ensured the required speed and many other target performance figures.

The main problems inherent in this layout were associated with the need to use new materials and technologies because in cruise flight the aircraft would soak at high temperatures caused by kinetic heating at high Mach numbers. Seeking to reduce the technical risks involved, the Tupolev OKB decided to restrict the new bomber's cruising speed to Mach 2.3, which was somewhat less than the competitors' figures.

Of course, the Tupolev OKB engineers also considered the variable-geometry layout. This layout offered certain advantages but imposed a weight penalty and complicated the design considerably.

One of the key requirements was long range over a complex mission profile involving air defence penetration at high altitude in supersonic mode (the so-called 'hi-hi-hi' mission profile) or at low altitude in subsonic mode (the 'hi-lo-hi' profile). Target approach was to be performed in subsonic cruise at optimum altitude. Another (albeit less critical) requirement was good field performance, allowing the bomber to operate from relatively short runways. Combining these virtues in a single aircraft was no simple task. An acceptable balance of subsonic and supersonic performance could only be obtained by utilising variable-geometry wings and unconventional compound engines operating as turbojets in supersonic mode and as turbofans in subsonic mode.

An aerodynamic comparison of the fixed-sweep and variable-geometry versions made when the optimum layout was being defined revealed that the VG version had a 20-50% better lift/drag ratio at subsonic speeds. In supersonic cruise with the wings at maximum sweep the L/D ratio was virtually equal to that of the fixed-sweep aircraft.

As already noted, an inherent major shortcoming of 'swing-wing' aircraft is the increase in empty weight caused by the massive wing pivots and actuators. Calculations showed that if this assembly made up more than 4% of the empty weight, the weight penalty negated all the advantages conferred on a heavy bomber by the variable-geometry wings. Assuming the powerplant was identical, a heavy VG aircraft possessed approximately 30-35% longer range at medium altitude and 10% longer range at low altitude than a conventional fixed-sweep aircraft when flying at subsonic speeds. In supersonic cruise at high altitude the range was about equal but again a 'swing-wing' heavy aircraft gained an advantage of approximately 15% at low altitude. VG aircraft also had better field performance.

As mentioned earlier, choosing the correct cruise Mach number was an important aspect of a supersonic heavy bomber's design process. Theoretical and aerodynamic research was undertaken to compare the range of a 'swing-wing' heavy aircraft at two different cruising speeds, Mach 2.2 and Mach 3.0. The results were unmistakable: reducing the cruising speed to Mach 2.2 enhanced range considerably thanks to a lower SFC and a better lift/drag ratio. Besides, the airframe of a bomber designed to cruise at Mach 3 would include a large proportion of titanium alloys, which increased unit costs and created additional technological problems, as titanium is difficult to machine and weld.

After assessing the merits and shortcomings of the various layouts the Tupolev OKB selected the tailless delta layout for its multi-mode strategic bomber. Several project versions were developed in 1970-72 under the provisional designation 'aircraft 160M' (aka *izdeliye* L); these were known as 'L-1', 'L-2' etc. The PD project of 'aircraft 160M' was completed in 1972 and submitted to the Soviet Air Force's Scientific & Technical Committee, a body which evaluated projects of new hardware before detail design could begin and the first metal could be cut. Concurrently the VVS assessed the Sukhoi T-4MS and Myasishchev M-18 projects; all three aircraft had been entered for the contest to create a new airborne strategic weapons system which was announced by MAP in 1972.

As might be imagined, the three projects were very different – the way a greyhound is different from a Great Dane or a German shepherd, though all of them are fast dogs. The different design practices and work styles of the three OKBs inevitably had their effect on the bombers' design (after all, the Sukhoi OKB was primarily a 'fighter maker' while the other two specialised in heavy aircraft from the outset). What they *did* have in common was the wish to use as many new design features as possible, providing they were technically viable; this was especially true for the Sukhoi and Myasishchev projects. Here it is worth quoting a book by Air Marshal (retired) Vasiliy V. Reshetnikov who was then in the rank of Colonel General and commanded the Soviet Air Force's long-range bomber arm (DA – **Dahl'**nyaya avi**ah**tsiya) in the 1970s:

'Since the situation with Tupolev was clear, the commission first paid a visit to Pavel Osipovich (ie, to P. O. Sukhoi's OKB-51 – Auth.). The aircraft he proposed overwhelmed you with its unconventional aerodynamic layout; it was almost a flying wing, with enough internal volume to accommodate both the engines, the weapons load and the fuel. What took me aback at first was the unusually thick airfoil of this giant wing with its hefty leading edge; that was not my idea of a supersonic aircraft. Feeling awkward, I asked Pavel Osipovich about it. However, he was prepared for this question; he showed me the design materials and demonstrated the test results obtained in TsAGI's supersonic wind tunnel. Gradually my doubts were dispelled and the project began to look quite realistic and appealing. The thick wings with the smoothly blended curves of its contours was probably Pavel Osipovich's brainchild, and he was eager to see it materialise on a large supersonic aircraft.

Equally interesting – and equally well thought out – was the project offered by Vladimir Mikhaïlovich Myasishchev. It was an elegant aircraft with a slender fuselage and a sleek, barracuda-like appearance that made it look much lighter than it really was. Oh, I would that it would have a chance to fly! As was his wont, Vladimir Mikhaïlovich, a brilliant designer with a wealth of experience in developing heavy combat aircraft, introduced many novel features into the aircraft's systems, not repeating what he had done before. The combat capabilities of this aircraft looked set to be among the highest in the world.'

In the autumn of 1972 the MAP's scientific and technical council convened to hear reports on the projects described above. It should be noted that, as mentioned earlier, Tupolev's 'aircraft 160M' had been developed in several versions differing in detail. The version eventually selected for participation in the tender featured double-delta wings – ie, with a kinked leading edge, as on the definitive production-standard Tu-144 (in-house designation '*izdeliye* 004'). (Note: The original Tu-144 prototype which first flew on 31st December 1968 (registration CCCP-68001, in-house designation '*izdeliye* 044') had ogival wings, ie, delta wings with an S-shaped leading edge; this aircraft showed disappointing performance and a complete redesign was undertaken.)

This proved fatal; the MAP and Air Force top brass knew the Tu-144 perfectly well, and when the '160M' was unveiled the similarity was all too obvious. As a result, the Tupolev OKB project was rejected on the grounds that it 'did not meet the specifications'. Commenting on the '160M' at a session of the Air Force's Scientific & Technical Committee, Col. Gen. Reshetnikov said that the Air Force was being offered a warmed-over airliner! The fact that an excessively high lift/drag ratio had been unintentionally quoted in the project documents certainly did not speak in the Tupolev bomber's favour either. Here's how Reshetnikov himself describes the episode in the aforementioned book:

'As we took our seats in a small conference room and examined the drawings and diagrams attached to a display stand, I was surprised to see the familiar lines of the Tu-144 supersonic airliner. Could it actually be the same aircraft? The Tu-144 fell short of its performance target, was beset by reliability problems, fuel-thirsty and difficult to operate. Moreover, there had been real disasters involving the type. The civil aviation would have no part of it...

(Note: While Reshetnikov correctly conveys the generally negative impression the '160M' project made on the military, he has mixed up the details a lot. The 'real disasters' bit must be a reference to the type's two fatal crashes; however, both of them (the crash of Tu-144 *sans suffixe* CCCP-77102 on 3rd June 1973 during a demonstration flight at the 30th Paris Air Show and the crash-landing of Tu-144D CCCP-77111 near Zhukovskiy on 25th May 1978 after an in-flight fire) occurred *after* the session where the '160M' was axed, which took place in 1972! Also, how would Reshetnikov know about operational difficulties with the Tu-144 in 1972 when revenue services did not begin until 1975?)

[...] Aleksey Andreyevich [Tupolev, the new General Designer] was not quite his usual self as he approached the stand, pointer in hand. His proposal boiled down to providing weapons bays for the bombs and missiles in the space between the engine packs occupy-

ing the fuselage undersurface. There is no need to relate Tupolev's discourse that followed; it was obvious that, weighed down by the offensive and defensive armament, this unsuccessful airliner-turned-bomber would be robbed of whatever structural strength reserves it had and all performance characteristics would drop.

About five or ten minutes later I rose and, cutting the lecture short, stated that we were not going to consider the project any longer because, even in revamped condition, an aircraft originally designed for carrying passengers would still retain some inherent properties which were absolutely unnecessary for a combat aircraft while still not meeting the demands applying to a strategic bomber.

Apparently Aleksey Andreyevich was prepared for this outcome. Without saying a word he turned towards the largest diagram pinned in the middle of the stand, grasped it and tore it down with a jerk. The sharp crack of heavy paper being rent asunder resounded in the complete silence. Then he faced me again and apologised, adding that he would invite us again when a new PD project would be ready.'

Here it would be appropriate to quote another passage from Reshetnikov's book explaining who in reality was behind the Tupolev OKB's failed attempt to foist the unwanted airliner on the military (in their opinion).

'However, Aleksey Andreyevich was not to blame in this incident. The development and construction of the supersonic airliner, the future Tu-144, was included in the five-year economic development plan and was under the auspices of the influential D. F. Ustinov (Marshal Dmitriy Fyodorovich Ustinov was then the Soviet Minister of Defence – *Auth*.) who regarded this mission as a personal responsibility – not so much to his country and people as to 'dear Leonid Il'yich' (Brezhnev, the head of state – *Auth*.) whom he literally worshipped – sometimes to the point of shamelessness...

Yet the supersonic passenger jet was apparently not making headway and, to the dismay of its curator, it looked like Brezhnev might be disappointed. It was then that Dmitriy Fyodorovich jumped at someone's bright idea to foist Aeroflot's 'bride in search of a wedding' on the military. After it had been rejected in bomber guise, Ustinov used the Military Industrial Commission (a standing committee on defence matters in the Soviet government – *Auth*.) to promote the aircraft to the Long-Range Aviation as a reconnaissance or ECM platform – or both. It was clear to me that these aircraft could not possibly work in concert with any bomber or missile carrier formations; likewise, I could not imagine them operating solo as 'Flying Dutchmen' in a war scenario, therefore I resolutely turned down the offer.

Naval Aviation Commander Aleksandr Alekseyevich Mironenko, with whom I had always worked in close co-operation, followed suit.

Nothing doing! Ustinov would not be put off that easily. He managed to persuade the Navy C-in-C [Admiral] S. G. Gorshkov who agreed to accept the Tu-144 for Naval Aviation service as a long-range maritime reconnaissance aircraft without consulting anyone on the matter. Mironenko rebelled against this decision, but the Commander-in-Chief would not hear or heed – the issue is decided, period. On learning of this I was extremely alarmed: if Mironenko had been pressured into taking the Tu-144, this meant I was going to be next. I made a phone call to Aleksandr Alekseyevich, urging him to take radical measures; I needn't have called because even without my urging Mironenko was giving his C-in-C a hard time. Finally Ustinov got wind of the mutiny and summoned Mironenko to his office. They had a long and heated discussion but eventually Mironenko succeeded in proving that Ustinov's ideas were unfounded. That was the last we heard of the Tu-144.'

Now we go back to the tender itself. Sukhoi's T-4MS (*izdeliye* 200) drew a very favourable reaction from the military and attracted a lot of attention. Myasishchev was less lucky; his project was highly commended (the commission stated it was carefully designed and met the Air Force's specifications) but nonetheless rejected because the recently reborn OKB lacked the necessary technological assets and manufacturing facilities for prototype construction. Plant No. 23 (the OKB's former experimental production facility) in Fili, an area in the western part of Moscow, had been transferred to Academician V. N. Chelomey's OKB and was now busy producing missiles; at the Myasishchev OKB's new premises in Zhukovskiy southeast of Moscow there was little more than a flight test facility.

(Interestingly, in the numerous press articles and books on Myasishchev OKB history that have appeared in recent years Myasishchev spokesmen and executives invariably call the M-18 the official winner of the 1972 tender. In reality apparently the winner was was never announced officially but appropriate remarks on the projects and the advisability of further proceeding with them were made in the protocols of the tendering commission. These protocols were followed by Council of Ministers directives and appropriate MAP orders which tasked the Tupolev OKB with the multi-mode strategic missile carrier programme. The minutes of the tendering commission and the final ruling of same are still classified, which puts both the Sukhoi and Myasishchev bureaux in a position to interpret the results of the tender as they see fit.

The long and short of it was that the Sukhoi OKB was declared the winner. This OKB had accumulated some experience of heavy aircraft construction and testing with the T-4 (*izdeliye* 100). However, to ensure construction of the T-4MS prototype the Kazan' aircraft factory (MAP factory No. 22, one of the Soviet Union's two principal factories producing heavy bombers) had to be assigned to the Sukhoi OKB. Nobody in the industry wanted to see it happen, except Sukhoi themselves. Besides, the OKB had its hands full, being tasked with developing the new T-10 advanced multi-role tactical fighter (which finally emerged as the famous Su-27 *Flanker*) and creating new versions of the Su-17M *Fitter-C* fighter-bomber and Su-24 *Fencer* tactical bomber. An involvement with heavy bombers jeopardised all these important programmes.

The final session of the tendering commission was summed up by Soviet Air Force C-in-C Air Marshal P. S. Kootakhov. 'Look here, – he said, – let's see it this way. Yes, the Sukhoi OKB's project is the best, we have given it due credit, but remember that the OKB is already heavily involved with the Su-27 fighter which we need badly. Therefore let's resolve the matter as follows: we'll acknowledge that the Sukhoi OKB has won the tender and then order it to transfer all project materials to the Tupolev OKB so that the latter can proceed further with the project...' It was similarly 'recommended' that the Myasishchev OKB hand over the M-18 project materials to Tupolev.

There you are. To quote a famous phrase from Kurt Vonnegut's book *Slaughterhouse Five*, so it goes. 'Frustrating' is too colourless a word to describe how it feels to win and then have your victory stolen in this fashion. And the more apposite words are unprintable ones. However, in those days there was little the OKBs could do to defend their progeny: government decisions had to be complied with.

Later, however, the Tupolev OKB rejected the T-4MS project (but not the M-18 project!) and continued working on an all-aluminium strategic strike aircraft with variable-geometry wings which ultimately became the Tu-160. One look at the present-day *Blackjack* is enough to see its M-18 origins, though the two aircraft have significant differences.

Chapter 2
Taking Shape

From M-18 to Tu-160

After this session where the crucial decisions concerning the Soviet new-generation strategic bomber were taken the Tupolev OKB started design work on the 'aircraft 160' (Tu-160) bomber/missile carrier (aka *izdeliye* 70) featuring variable-geometry wings. Fixed-sweep configurations were not considered any more. That same year (in 1972) OKB-156 launched a large-scale R&D programme aimed at optimising the aerodynamic layout, powerplant and various parameters of the future aircraft, selecting the proper structural materials and technologies and integrating the avionics and armament into a close-knit complex. The Tupolev OKB's partners in this effort included TsAGI, the State Research Institute of Aircraft Systems (GosNII AS – *Gosoodarstvennyy naoochno-issledovatel'-skiy institoot aviatsionnykh sistem*), the Flight Research Institute named after Mikhail M. Gromov (LII – **Lyot**no-is**sled**ovatel'skiy insti**toot**), the All-Union Research Institute of Aviation Materials (VIAM – *Vsesoyooznyy institoot aviatsionnykh materiahlov*), the Research Institute of Aviation Hardware (NIAT – *Naoochnyy institoot aviatsionnoy tekhniki*), the Moscow Institute of Electronics (MIREA – *Moskovskiy institoot rahdioelektronnoy apparatoory*), MKB Raduga, NPO Trood (Labour; NPO = *naoochno-proizvodstvennoye obyedineniye* – scientific and production association), NPO Elektroavtomatika in Leningrad etc. Assistance was also provided by the Soviet Air Force's R&D establishments.

All in all, nearly 800 enterprises and institutions working in various areas were involved in the Tu-160 programme which proceeded under the overall supervision of General Designer Aleksey A. Tupolev. The actual design effort and subsequent prototype construction were directly led by V. I. Bliznyuk, the Tu-160 project chief, and his aides L. N. Bazenkov and A. L. Pookhov.

A special responsibility rested with GosNII AS, the main design institute which determined the general outlook of the new weapons system. Since 1969 the institute had been working on the technical concept of the new multi-mode strategic strike aircraft, analysing the main parameters and assessing the efficiency of the various project versions proposed by the competing design bureaux. At the same time the subdivisions of GosNII AS founded the bomber's weapons fit and avionics/equipment complement. The work was performed by B. P. Toporov, O. S. Korotin, G. K. Kolosov, A. M. Zherebin, Yu. A. Volkov and other employees of the institute.

After the tendering commission ruled that the Tupolev OKB should take over the new-generation strategic bomber programme, GosNII AS started developing the Tu-160's technical outlook and performing combat efficiency studies. As a result, the ultimate choice of the VG layout for the Tu-160 was largely influenced by the expert opinion of GosNII AS founded mainly on calculations and analysis by the institute's Section 1.

From an early stage GosNII AS insisted that multiplex data exchange channels be used and all the latest avionics be integrated into the bomber's mission avionics suite. As a result, the latter came to include a separate subsystem called Missile Control System (MCS) – a phenomenon brought about by the new generation of air-to-surface missiles which required a large amount of data to be prepared on board the aircraft and downloaded to the missile's computer prior to launch. The need was acknowledged to develop a flight data preparation system allowing the complete mission to be prepared on the ground and uploaded to the aircraft's mainframe computer – a task which was solved in due course. Throughout the design effort GosNII AS worked in close co-operation with departments of the relevant organisations headed by L. N. Bazenkov (Tupolev OKB), O. N. Nekrasov and V. F. Khoodov (MIREA).

The institute started scientific and technical support of the Tu-160's PD project development in 1972. Anticipating the large amount of work to be done on heavy bombers, in August 1974 Academician Yevgeniy A. Fedosov, the head of GosNII AS, drew resources from several of the institute's departments and laboratories to create a new division, Section 14. Headed by V. I. Chervin, this department was directly responsible for the development of airborne strategic strike systems.

Due to the fact that this area of work was led by Academician Fedosov himself, Section 14 quickly grew from a 60-man team to the largest research department within the GosNII AS structure, with a 290-strong staff and the best-equipped laboratories. The Tu-160 programme was the section's main work project. By then Section 14 had moved into a new building – specially built with the Tu-160 programme in mind. The staff started outfitting the new laboratory halls and mastering the new equipment; the work often proceeded in three shifts due to the pressure of time.

For the first time ever the staff of Section 14 had to deal with such challenges as exactly aligning the axes of a cruise missile's inertial navigation system (INS) with those of the carrier aircraft's INS, downloading a digital route map from the aircraft to the missile via datalink, preparing the missile's automatic flight control system for launch and controlling the launch of missiles from internal bays and external hardpoints. Together with the institute's Section 5 headed by K. A. Sarychev, the algorithms of an autonomous correlation-extreme missile guidance system were developed, flight test results were analysed and so on.

A research effort undertaken by Section 5 under the direction of V. I. Chervin (that is, before Section 14 was formed) resulted in a proposal concerning reasonably simple and affordable long-range cruise missiles. Actually these ideas had crystallised long before the USA had begun developing strategic cruise missiles, but the Soviet military leaders were not interested at the time. Serious research in this field at GosNII AS began only when alarming news started coming from the USA – and then, figuratively speaking, the Soviet researchers had to run hell-for-leather, working hard to complete development within the shortest possible time. This certainly took a maximum of effort.

The new multi-mode strategic strike aircraft programme enjoyed top priority with MAP at the time; conferences on this subject were held every now and again by Minister of Aircraft Industry Ivan S. Silayev and special state prizes were awarded to those 'subcontractors' which were doing well. Everyone in the industry was aware that the success (or failure) of the strategic weapons system built around the Tu-160 depended heavily on the work done by GosNII AS specialists. It may as well be said now that all R&D work on the

Valeriy Ivanovich Bliznyuk became the Tu-160's chief project engineer soon after the commencement of full-scale development.

weapons system's components – the Tu-160 carrier aircraft, the cruise missile and the Sproot-SM (Squid-SM) target data preparation system – was completed and the results delivered to the Air Force, the Tupolev OKB, MKB Raduga and other enterprises right on schedule.

The Central Aerodynamics & Hydrodynamics Institute, which traditionally maintained close ties with the Tupolev OKB, played a vital role at the early stage of the 'swing-wing' Tu-160's development. The honour goes first and foremost to TsAGI employees G. S. Büschgens and Gheorgiy P. Svischchev who were awarded the State Prize in 1975 in recognition of their contribution to the programme.

Now let's get down to the meat and potatoes. What did the new Tupolev bomber look like from a design standpoint? The choice of general arrangement and aerodynamic features was closely tied to structural design and manufacturing technology issues. Even though the 'militarised' Tu-144 concept had been rejected, some of the *Charger*'s design features found their way to the Tu-160. These included elements of the integral (blended wing/body) layout where the large leading-edge root extensions (LERXes) were manufactured as a single entity with the fuselage. This allowed the engineers to kill three birds with one stone, making the structure very lightweight for its size while creating additional lift and providing large internal volumes for weapons and fuel. (Speaking of which, the BWB technology was much more apparent on the Tu-160 with its smoothly blended contours than on the Tu-144 which basically had a circular-section fuselage sitting on top of sharp-edged LERXes.) As a result, the Tu-160 was 50% heavier than the similarly sized Tu-95 *Bear* – thanks to a higher payload.

Generally the wing design of the 'aircraft 160' (*izdeliye* 70) was borrowed from Tupolev's previous VG bomber, the Tu-22M *Backfire* (*izdeliye* 45). The movable outer wings, wing pivots and actuators were similar to those of the *Backfire*. However, the new aircraft was a lot larger (the maximum all-up weight was three times higher!); this and the much higher aerodynamic loads required major structural changes and the provision of more powerful actuators. The outer wings were of five-spar construction, each featuring seven chemically milled skin panels with integral stiffeners (three on the upper surface and four on the undersurface) and only six ribs. They were hinged to a hefty wing pivot box carry-through unit – a large welded titanium structure which the entire airframe was effectively built around; it absorbed all the principal loads acting on the aircraft. Manufacturing such a large titanium structure was made possible by an unusual technology – laser welding in an inert gas environment; this technology remains unique to this day and can be justly regarded as Russian know-how.

To maximise the lift/drag ratio at all possible wing sweep angles the OKB devised a system of hinged flaps sealing the wing glove joint. Later, on production Tu-160s, these gave way to a feature unique among VG aircraft – the inboard ends of the wing flaps hinged upward as the wings pivoted to form prominent boundary layer fences. In addition to optimising the airflow near the joint, these 'pop-up' fences obviated the need to seal the wing glove joint (which is always difficult). A large amount of wind tunnel testing was undertaken at TsAGI to verify the chosen aerodynamic features, using 11 scale models. The tests showed that the engineers had succeeded in obtaining a maximum lift/drag ratio of 18.5-19 in subsonic cruise and more than 6.0 in supersonic cruise.

Still, designing variable geometry wings and wing pivots for such a heavy aircraft proved to be a major challenge. Using VG wings on a strategic bomber required a qualitatively new level of manufacturing technology. This brought into existence a special State Programme of New Metallurgic Technology Development co-ordinated by Pyotr V. Dement'yev, the new Minister of Aircraft Industry.

One area where the engineers found themselves in a tight spot – literally – was the tail unit. All-movable tail surfaces had been chosen for the Tu-160. The big problem was how to accommodate the hefty upper fin and stabiliser pivots and extremely powerful electrohydraulic actuators inside the relatively thin airfoils (to work the large control surfaces the actuators were required to develop a force of around 7,000 kgf (15,430 lbf). It has to be said that originally a T-tail with slab (ie, all-movable) stabilisers on top of a conventional fin/rudder assembly was envisaged. An alternative configuration featured cruciform tail surfaces with a rudder divided into upper and lower sections (as, eg, on the Rockwell International B-1). Eventually, however, an unconventional arrangement with mid-set slab stabilisers and a vertical tail divided into two portions (with an all-movable upper half) was chosen for the Tu-160; this left enough room for the actuators inside the fixed lower portion of the fin which also carried the horizontal tail.

Another issue which took a long time to resolve was the positioning of the weapons bays with respect to the fuselage length. At first the engineers wanted to place the two weapons bays side by side in the centre fuselage; this minimised centre of gravity (CG) travel when the weapons were dropped or launched. On the minus side, this increased the fuselage cross-section area (and hence drag) and complicated engine nacelle design. Hence the side-by-side arrangement was abandoned in favour of tandem bays.

Generally speaking, the Tu-160's dimensions were kept to a minimum thanks to a well thought-out structural layout. For instance, to minimise fuselage cross-section area the nosewheel well was placed aft of the flight deck rather than below it, as on the B-1. The main gear units had a complex retraction sequence, contracting during retraction to fold into the smallest possible space. Drag reduction was also helped by the fuselage's high fineness ratio and carefully streamlined forward fuselage contours with a sharply raked windshield (initially the engineers planned to provide the Tu-160 with a drooping nose *à la* Tu-144 and Sukhoi T-4 but then dropped the idea). Besides ensuring the required speed and range, these measures conferred a sleek look on the bomber. (As one Russian female newspaper journalist later remarked when writing about the Tu-160, *'appearances are deceptive; the graceful forms of this aircraft belie the fact that it is an instrument of death and destruction'*.)

The powerplant selected originally consisted of four Kuznetsov NK-25 three-spool afterburning turbofans each rated at 14,300 kgp (31,525 lb st) dry and 25,000 kgp (55,115 lb st) reheat. This engine (manufacturer's designation *izdeliye* Ye) powered the Tu-22M3 *Backfire-C*. The NK-25 had the required thrust but the SFC was way too high, making intercontinental range impossible even if the bomber had ideally refined aerodynamics. It was then that OKB-276 started work on a new three-spool turbofan, the NK-32. Possessing an identical thrust in full afterburner and a slightly reduced dry thrust of 13,000 kgp (28,660 lb st), the new engine was expected to have an agreeable SFC of 0.72-0.73 kg/kgp hr in subsonic mode and 1.7 kg/kgp hr in supersonic mode. The NK-32 had considerable structural commonality with the production NK-25, which made it a realistic option.

Unlike the Tu-144, the Tu-160's powerplant and everything that went with it – the engines proper, the engine nacelles, the variable supersonic air intakes and the placement of the engines – was designed with the airframers and propulsion engineers working in close co-operation from the start. This allowed many of the deficiencies that plagued the luckless SST's powerplant to be designed out of the bomber at an early stage.

Earlier, the Myasishchev OKB had selected a powerplant arrangement for the M-18 similar to that of the B-1 (the engines were housed side by side in nacelles adhering directly to the wing undersurface close to the wing pivots). Now the M-18 formed the basis for the 'aircraft 160' (izdeliye 70) project; nevertheless, the Tupolev OKB decided to pursue its own approach and see if something better could be devised. A major wind tunnel effort was undertaken jointly with TsAGI, involving no fewer than 14 models of various powerplant configurations.

The layout used on the 'first-generation' Tu-144 (izdeliye 044) was tried first, with all four engines in a common nacelle placed under the wing centre section trailing edge. This allowed the slanting shock waves generated in supersonic cruise to be utilised for improving the lift/drag ratio. However, the arrangement also led to excessive loss of inlet pressure in the long inlet ducts; moreover, individual adjustment of the intake ramps could have an adverse effect on the neighbouring intakes. Finally, having the engines mounted so close together increased the risk of the neighbouring engines being put out of action, should one of the engines suffer an uncontained failure or fire. (This is obviously one of the reasons why the engines were moved apart in pairs on the production 'second-generation' Tu-144, izdeliye 004.)

The most unusual configurations were proposed, including paired engines located one above the other à la BAC Lightning, with horizontal intake ramps; another version had three engines breathing through circular intakes with centrebody shock cones. The 'double-decker nacelle' version even reached the full-scale mock-up stage; the inlet ducts of the two engines curved around the wing pivot box carry-through structure, passing above and below it. This arrangement offered both the lowest possible drag and the lowest radar cross section (RCS), ie, maximum 'stealth' – an all-important quality for a strategic bomber. However, technological problems and doubts as to the combat survivability of the Lightning-style engine packs caused this version to be abandoned.

Eventually the engineers reverted to the layout featured in the M-18 project, with side-by-side pairs of engines and two-dimensional variable air intakes featuring vertical ramps. A similar intake design had undergone comprehensive testing on the Tu-144 (save for the fact that the Charger had horizontal intake ramps). The main reason why the nacelles were placed under the extremities of the integral wing centre section was the need to free the centreline for the weapons bays which had to be as close to the CG as possible.

The bomber was to make large-scale use of the latest structural materials; titanium alloys made up 38% of the airframe weight, with aluminium alloys, high-strength steels and composites making up 58%, 15% and 3% respectively. Once the general arrangement had been frozen the OKB concentrated on detail design.

The Tu-160 was the first Soviet heavy aircraft to feature a fly-by-wire (FBW) control system with no mechanical connection between the controls and the control surfaces. This allowed the aircraft to be electronically stabilised in flight with the CG in neutral position. Another 'first' in Soviet bomber design practice was the provision of a fighter-type 'joystick' instead of the usual control column. These measures increased the aircraft's range (due to lower drag at optimum flight attitudes), improved controllability and helped reduce crew fatigue in complex flight situations.

As already mentioned, the Tupolev OKB teamed with GosNII AS and other establishments to develop the most effective missile armament system for the Tu-160. Since the geopolitical and military situation in the years to come was unforeseeable, the aircraft's weapons complement was to be determined by its multi-role status. The aircraft was intended to carry ultra-long-range, long-range and medium-range cruise missiles, guided and unguided short-range weapons (ie, 'smart' and 'dumb' bombs); additionally, air-to-air missiles would be provided for self-defence. Priority was given to weapons which could destroy targets (including those with a small radar signature) without requiring the bomber to come within range of the enemy's air defences and were carried internally. The bomber's avionics suite was required to tackle navigation tasks and ensure accurate delivery of a wide range of weapons.

Originally the Tu-160's principal missile armament was to consist of either two Kh-45 long-range missiles (one in each weapons bay) or 24 Kh-15 short-range cruise missiles carried on four MKU-6-1 rotary launchers (MKU = **mno**gopozits**on**naya kata**pool't**naya oosta**nov**ka – lit. 'multi-position ejector unit',

Yevgeniy Fedosov, head of GosNII AS, supervised the development of the Tu-160's mission avionics and weapons system.

ie, ejector rack). These armament options determined the size of the weapons bays; the large size of the Kh-45 measuring 10.8 m (35 ft 6 in) in length and 1.92 m (6 ft 3½ in) in height with fins folded was the decisive factor. The missile had a launch weight of 4.5 tons (9,920 lb) and a design range of 1,000 km (621 miles); different sources quote a maximum speed of 7,000 to 9,000 km/h (4,350-5,590 mph; 3,780-4,865 kts). Each of the two weapons bays was about equal in volume to the Tu-95 Bear-A's bomb bay. In addition to low-level supersonic missiles, aircraft in the Tu-160's class were to be armed with low-level subsonic cruise missiles featuring a terrain-following correlation navigation system – development of which had yet to begin.

Initially the VVS insisted that, in keeping with previous bomber design traditions, the Tu-160 be equipped with a tail barbette mounting a 30-mm (1.18 calibre) Gryazev/Shipoonov GSh-6-30 six-barrel Gatling cannon. Pretty soon, however, the Tupolev OKB persuaded the military to dispense with cannon armament in favour of a more capable ECM system, using the weight saving and extra space freed by the deletion of the tail barbette. In addition, a specialised ECM version designated Tu-160PP (posta**nov**schchik po**mekh** – ECM aircraft) was proposed; this aircraft would provide cover for large bomber formations.

Another departure from previous design practices was the incorporation of stealth technology into the Tu-160's airframe and powerplant. The engineers took great pains to minimise the bomber's RCS and heat signature.

As already mentioned, the Tu-160's mission avionics suite was designed to work

29

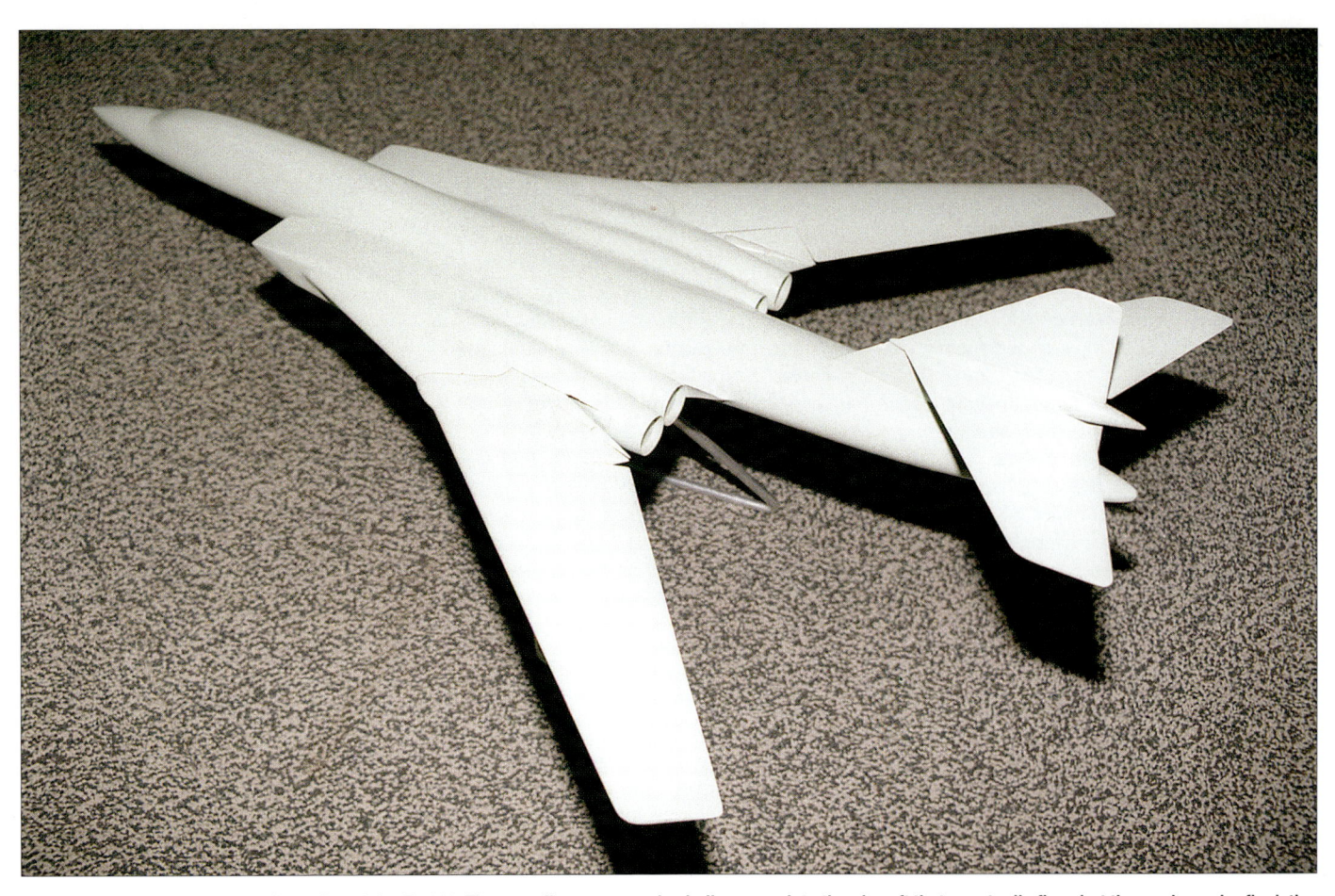

Above and below: An early configuration of the Tu-160. The overall appearance is similar enough to the aircraft that eventually flew, but the engine pairs flank the fuselage, breathing through raked lateral air intakes similar to those of the Tu-22M3.

Two more views of the PD project model of the Tu-160. Note that the cruciform tail unit did not yet feature a fin fillet at this stage.

31

Above and left: The Tu-160's Kuznetsov NK-32 engine was put through its paces on this Tu-142LL engine testbed operated by the Flight Research Institute (LII) – a converted Tu-142M ASW aircraft (c/n 4243). It is seen here at LII's airfield in Zhukovskiy in March 1991. The NK-32 was housed in a large nacelle which was semi-recessed in the former weapons bay to give adequate ground clearance; it was lowered in flight to take the efflux of the development engine well clear of the fuselage.

Bottom: This close-up of the NK-32's nacelle in the stowed position emphasises the engine's large dimensions. A circular cover retracting forward into the fairing visible ahead of the air intake protected the engine against foreign object damage during take-off and landing. Here the Tu-142LL is parked over a special trench with an exhaust duct allowing the engine to be lowered for ground runs.

with several cruise missile types. At General Designer Aleksey A. Tupolev's request prototype avionics integration took place at GosNII AS where a simulation complex utilising actual hardware components was specially built for the purpose. The high quality of the test and debugging work speeded up the commencement of flight tests a lot. It should be noted that later, in 1981, the institute built the world's largest pressure chamber measuring 40.0 x 18.0 x 9.8 m (131 ft 2¾ in x 59 ft ¾ in x 32 ft 1¾ in) specially for testing the bomber's avionics and equipment in simulated high-altitude conditions.

The sophisticated and highly complex weapons control system necessitated large-scale use of computers. As already mentioned, at the insistence of GosNII AS the WCS had provisions for using multiplex data exchange channels and a separate missile control system (MCS) for downloading a large amount of data to the missile prior to launch.

Faced with the ever-growing lethality of the potential adversary's air defence systems, in the 1970s the aircraft designers had to focus on the combat survivability aspect of strike aircraft. Hence a special new subdivision headed by O. S. Korotin was formed at GosNII AS to develop the structure and assess the efficiency of defensive avionics systems; it was transformed into a laboratory in 1984. This team analysed the effectiveness of the Tu-160's Baikal electronic support measures (ESM) suite and founded the choice of its components.

On 26th June 1974 the Council of Ministers issued a directive tasking the Tupolev OKB with developing the Tu-160 multi-role strategic bomber/missile strike aircraft powered by four NK-32 engines. Yet another CofM directive (No. 1040-348) appeared on 19th December 1975; this was a 'follow-up' document specifying more accurate main performance figures. In accordance with these documents the bomber's effective range in subsonic cruise mode with 9,000 kg (19,840 lb) of ordnance – ie, two Kh-45 missiles – was to be 14,000-16,000 km (8,695-9,940 miles); range over a 'hi-lo-hi' mission profile including a 2,000-km (1,240-mile) stretch of low/ultra-low-level flight (at 50-200 m/165-660 ft) or in supersonic cruise mode was specified as 12,000-13,000 km (7,450-8,075 miles). Maximum speed was required to be 2,300-2,500 km/h (1,430-1,550 mph; 1,243-1,351 kts) at high altitude and 1,000 km/h (620 mph; 540 kts) at low altitude. The service ceiling was stated as not less than 18,000-20,000 m (59,055-65,620 ft).

Above: The Tu-160's Obzor-K radar was tested by LNPO Leninets on the SL-18V avionics testbed converted from IL-18V CCCP-75786 (c/n 181003905). It is seen here in the original livery worn by Aeroflot IL-18s until the mid-1970s and with a fixed Kh-45 acquisition round under the forward fuselage.

The SL-18V at a later date in 1973-standard Aeroflot colours; note the pylon for the acquisition round.

The aircraft was to have a normal weapons load of 9,000 kg and a 40,000-kg (88,180-lb) maximum weapons load. The following missile armament options were set forth: two Kh-45M missiles; 24 Kh-15 missiles; ten or twelve Kh-15M missiles; and ten or twelve Kh-55 subsonic cruise missiles. The aircraft was to be capable of delivering conventional and nuclear free-fall bombs, as well as laser-guided and TV-guided 'smart bombs'.

With the official state order for the development of the new bomber finally placed and the main general arrangement and systems issues resolved, OKB-156 set to work on the advanced development project. At this stage the Kh-45 missile was relegated to second place among the Tu-160's weapons options, if not yet eliminated from the list altogether. Jointly with MKB Raduga the Tupolev OKB submitted technical proposals for a cruise missile to be used against ground targets and surface ships with a low radar signature; the missile was to be developed in standard (with a conventional warhead) and strategic (nuclear-tipped) versions designated Kh-55 and Kh-55SM respectively. At first the chiefs of MAP and the Air Force decided not to go ahead with the strategic version. However, they changed their minds in 1976 when it became clear that that the USA was rapidly developing the strategic ALCM-B (Boeing AGM-86B) air-launched cruise missile, and work on the Kh-55SM was resumed. The new missile was longer than the Kh-15, necessitating development of a new rotary launcher.

Here we will turn to Col. Gen. Vasiliy V. Reshetnikov's memoirs again. They bear testimony to the hard time the Tupolev OKB had when it came to defending the ADP of the 'aircraft 160' (*izdeliye* 70).

'In the Oval Hall of the Tupolev OKB Aleksey Andreyevich, looking very composed and solemn, was presenting the preliminary design project (*sic* – Auth.) of a new bomber called Tu-160.

For a minute or two we were silent, taking in the estimated performance presented in the tables and graphs, studying the technological breakdown of the airframe and mentally assembling it into a complete aircraft. The unfamiliar looks of the bomber were stern and forbidding, even though it bore a certain outward resemblance to the American B-1...

The reports delivered to us were so detailed as to seemingly preclude any unclarity. Yet both the General Designer and his aides were literally swamped in questions. It was plain to see that in the estimates the engineers were taking the aircraft to the limit. (Sic; what Reshetnikov meant to say is that the estimated performance figures obviously looked over-optimistic – Auth.) However, in real life all performance figures would inevitably slip dramatically. What then? Where would the real Mach limit be? What would the actual range be? What if the L/D ratio turned out to be lower than advertised? Would not the variable geometry wings make the aircraft overweight and sluggish? The questions kept coming, giving rise to new ones, and the answers were slow in coming.

My work group, which later transformed itself into the Tu-160's mock-up review commission and then into the State commission, spent a lot of time working at the OKB in frequent sessions. Weight figures were updated almost daily; the goddamn weight kept rising like a doomed patient's fever, a ton here and a ton there adding up to dozens, while the subcontractors responsible for the equipment and armament (or, at any rate, most of them) kept shamelessly cramming their obsolete, overweight goodies into the aircraft without a word of apology. And there was no stopping them because they had no competition. Balancing like a tightrope walker over an abyss, TsAGI worked hard to save the estimated performance, calculating the aerodynamics over and over again. Yet the recommendations they gave were to no avail because the weight kept rising still further.'

Still, the advance development project was eventually completed in 1976 and the following year a full-scale wooden mock-up was built at the OKB's experimental production facility, MMZ No. 156 'Opyt' in Radio Street. (MMZ = *Moskovskiy mashinostroitel'nyy zavod* – Moscow Machinery Plant number such and such; the name translates as either 'experiment' or 'experience'. This was also the unclassified name of the Tupolev OKB.) In 1977 the ADP and the mock-up were duly approved by 'the customer', albeit the Air Force pointed out a few bugs which had to be eliminated.

OKB-156 engineers Gheorgiy Alekseyevich Cheryomukhin, V. V. Soolimenkov, D. I. Gapeyev, Ya. A. Livshitz, N. T. Kozlov, V. M. Razoomikhin, V. A. Vishnevskiy, R. A. Yengoolatov, A. K. Yashookov, V. I. Korneyev, Ye. K. Moiseyev, A. V. Babochkin, Vladimir Mikhaïlovich Vool', A. S. Smirnov, V. P. Vorkin, Yu. S. Gorbanenko and many other engineers and workers of MMZ No. 156 contributed a lot to the Tu-160's development programme.

As per ADP documents the Tu-160 had an estimated take-off weight of 260 tons (573,190 lb) and an operating empty weight of 103 tons (227,070 lb); the fuel load was 148 tons (326,280 lb) and the normal weapons load was 9 tons (19,840 lb). The aircraft was slightly larger than its American counterpart and look-alike, the B-1A.

Later the weapons range was somewhat narrowed; the OKB chose not to use the Kh-45 missile, leaving the Tu-160 with either air-launched versions of the Kh-55 on two six-round rotary launchers or Kh-15s on four rotary launchers (plus various bombs). Eventually, however, the missile armament was restricted to a single type, the Kh-55, of which twelve were carried internally.

In 1977 the Kuznetsov OKB started design work on the NK-32 turbofan which received the manufacturer's designation *izdeliye* R. A flight-cleared engine entered flight test at LII in 1980 on the second of two Tu-142LL engine testbeds, a converted Tu-142M *Bear-F Mod 2* ASW aircraft (c/n 4243); the NK-32 was located in a large semi-retractable nacelle under the former weapons bay.

Concurrently the All-Union Electronics Research Institute (VNIIRA – *Vsesoyooznyy naoochno-issledovatel'skiy institoot rahdioelektroniki*) in Leningrad, aka LNPO Leninets (Leninist), proceeded with the development of the Tu-160's Obzor-K (Perspective-K) navigation/attack radar. This establishment, a division of the Ministry of Electronic Industry (MRP – *Ministerstvo rahdioelektronnoy promyshlennosti*), was one of the Soviet Union's leading avionics houses; it is now known as the Leninets Holding Company. In the late 1970s an Il'yushin IL-18V *Coot* turboprop airliner (CCCP-75786, c/n 181003905) was converted into the SL-18V testbed; the Obzor-K radar enclosed by a long conical radome was installed on a special adapter supplanting the IL-18's RPSN-1 *Emblema* (Emblem) weather radar. The SL-18V operated from the flight test centre in the town of Pushkin near Leningrad, sometimes venturing out over international waters where it was intercepted and photographed by Swedish Air Force fighters.

In 1974, when the Soviet government had only just issued the official order for the development of the new strategic bomber, the tests of the B-1A (to which the Tu-160 came to be quite similar at first glance) were already underway in the USA. Illustrated here is the second prototype, 74-0160 (c/n 3), as originally flown.

Chapter 3

Tests and Production

The *Blackjack* Becomes Reality

Tu-160 (*izdeliye* 70) prototypes

Construction of the first three Tu-160 airframes began at MMZ No. 156 in 1977 (again!). The work proceeded in close co-operation with the Kazan' aircraft factory named after Sergey Petrovich Gorboonov (MAP factory No. 22) which was concurrently gearing up for full-scale production of the new bomber. (Since the late 1980s the factory is known as KAPO – *Ka**zahn**skoye aviatsi**on**noye proiz**vod**stvennoye obyedi**nen**iye **im**eni S. P. Gorbo**o**nova*, Kazan' Aircraft Production Association.) It should be noted that in the Soviet Union it was fairly common practice to launch series production long before the trials programme had been completed – unless the aircraft was obviously a dead duck. In this case, however, preparations for series production began *before the Tu-160 had ever flown*! This says a lot for the faith the leaders of the Soviet state, Air Force and aviation industry had in the new bomber; or, equally probably, this was a measure of the need to create and field a Soviet 'anti-B-1' in response to the new threat from the United States.

The first flying prototype was known in-house as *izdeliye* 70-00, which gave rise to the OKB slang appellation *noolyovka* ('Aircraft Zilch'). The Air Force, however, referred to it as *izdeliye* 70-01 – which was more logical, as there was no *izdeliye* 70-01 in the OKB enumeration. *Izdeliye* 70-02 was the static test airframe, while *izdeliye* 70-03 was the second prototype – or, alternatively, it may be regarded as the sole pre-production aircraft.

Construction of a first prototype is always a protracted affair, since the aircraft is virtually hand-crafted; this is especially true for an aircraft of this complexity. In the summer of 1980 the partially completed prototype was delivered by road to the now-famous LII airfield in the town of Zhukovskiy south of Moscow where OKB-156 had its flight test facility. Ground systems checks of the still-incomplete aircraft began on 22nd October 1980. In January 1981 the prototype was finally completed but still far from ready for flight tests. The ground test and systems refining stage continued until November, lasting nearly ten months.

On 14th November 1981 the unpainted Tu-160 prototype – it never did receive a coat of paint during its flying career – moved under its own power for the first time, controlled by a test crew under OKB test pilot Boris Ivanovich Veremey. By the end of the month the aircraft had made three high-speed taxi runs. Finally, on 18th December Veremey took the 275-ton (606,260-lb) bomber aloft for its maiden flight. The event took place a few days before Leonid I. Brezhnev's 75th birthday and was universally regarded as the Tupolev OKB's (and the Soviet aircraft industry's at large) gift to the Soviet leader. Well, before the reader denounces the Tupolev OKB for 'toadying', it should be noted that timing major achievements to coincide with Communist Party congresses and so on was common in the Cold War-era Soviet Union.

The aircraft was captained by project test pilot Boris I. Veremey, assisted by co-pilot Sergey T. Agapov and navigators Mikhail M. Kozel and Anatoliy Yeriomenko. (The double quota of navigators (***shtoor**man* in Russian) is explained by Russian terminological subtleties. One of the two was really the navigator (***shtoor**man-navi**gah**tor*) while the other was not a navigator at all but the weapons systems operator (WSO, or ***shtoor**man-ope**rah**tor*). He

The unpainted first prototype Tu-160 seen in December 1978 soon after its maiden flight. Note the ogival radome tipped by a pitot.

Above: The first published photo of the Tu-160 parked next to two Tu-144s at the LII airfield. The picture was covertly taken by a passenger on an Aeroflot aircraft bound for Moscow-Vnukovo.

was thus called because the WSO had to know a lot about navigation to aim his weapons accurately; it was the same story on the Tu-22M.) We'll let Boris Veremey tell the story:

'The bomber looked set to take to the air on 19th December. That was Leonid Il'yich Brezhnev's birthday. It is an established fact that the Secretary General [of the Communist Party] paid much attention to the progress of the Armed Forces, which is why the original intention was to make him a gift in the form of a new bomber's first flight that same day. However, we had received no explicit instructions to make the Tu-160's maiden flight on 19th December, and the flight took place a day earlier. I still remember the 27-minute first flight of 'double zero' (ie, izdeliye 70-00 or noolyovka – Auth.) in the vicinity of the airfield; we took off, climbed to 2,000 m (6,560 ft), proceeded to the test flying area 150-220 km (93-136 miles) away, then came back and landed. The 'one sixty' had received its baptism.

The flight was carefully analysed and our comments were checked out on a ground test rig. For the first time in the OKB's history a hydromechanical test rig was used to check and refine the aircraft's systems, as is customary in the world's top aircraft manufacturing companies. (The Boeing Company calls such rigs 'iron birds' – Auth.) A special test department headed by project engineer Anatoliy Yashukov was formed (for the Tu-160 programme – Auth.). We made about 20 flights.

From the 13th flight onwards we switched places with Sergey Agapov – he flew the aircraft from the captain's seat in order to get his own impression of the bomber's handling.

The aircraft behaved predictably; the data obtained in the first flight matched the ground simulation results. Changes were introduced into the flight control system right away at the pilots' suggestion; the Tu-160 had a fly-by-wire control system which made such changes possible. [...] Besides yours truly, test pilots Sergey Agapov, Vladimir Smirnov and Nayil' Sattarov also took part in the trials.

Presently it was time to unveil the aircraft to the leaders of the nation. I remember the day in early 1983 when Marshal of the Soviet Union Dmitriy Fyodorovich Ustinov came to the airfield (in Zhukovskiy – Auth.). On seeing the Tu-160 he was overjoyed and started scurrying around the aircraft; as he later told us, he felt twenty years younger all at once. The Marshal postponed all the intended meetings at the OKB, spending all the time he had near the aircraft. He was very happy that we now had such a bomber.'

Yet long before that day news of the latest Soviet bomber had reached the West. Those who have attended the biennial Moscow airshows have probably paid attention to the large wall-less sheds along the aircraft parking areas on both sides of Zhukovskiy's main runway, but not many know their true purpose. The objective of these structures was to conceal sensitive new hardware from US surveillance satellites without occupying valuable hangar space which might be needed for maintenance or modification work. On 25th November 1981, one day before the second high-speed run, the prototype had been towed into the open from the shelter of such a shed and parked next to two of the four Tu-144D SSTs operated by the Tupolev OKB. As luck would have it, the very same day a Snooping Satellite passed over Zhukovskiy and photographed the scene.

Knowing the dimensions of the Tu-144 and using it as a reference point, American experts were able to calculate the dimensions of the new bomber with great accuracy. Henceforward the CIA maintained a close watch on the Tu-160's development process. Not knowing the exact designation, NATO's Air Standards Co-ordinating Committee (ASCC) initially allocated the provisional reporting name Ram-P to the new bomber. 'Ram' stood for Ramenskoye, the name of the neighbouring town which for many years was erroneously believed to be the name of the flight test centre in Zhukovskiy. (By comparison, the Sukhoi Su-27 Flanker heavy fighter, the Mikoyan MiG-29 Fulcrum light fighter and the Myasishchev M-17 Stratosfera/Mystic high-altitude aircraft were initially known to the West as Ram-K, Ram-L and Ram-M respectively.)

An artist's impression of the Tu-160 published in the Western press at about the same time as the bomber was shown to US Secretary of Defense Frank C. Carlucci. Note the inset rudder and elevators.

Above and below: The first prototype on short finals to Zhukovskiy after a test flight. These photos illustrate well the tandem weapons bays, the wing high-lift devices and the somewhat 'patchwork' appearance of the aircraft.

Above: The first Tu-160 makes a high-speed flypast with the wings at maximum sweep during one of its airshow appearances. Note the different shades of the skin panels which are made of electrochemically coated duralumin and titanium.

This view shows clearly the characteristic two-piece wing fences which pop up when the wings are fully swept back and the long pointed fin bullet fairing.

Later the reporting name was changed to *Blackjack* (in the 'B for Bomber' series); oddly enough, as already noted, this name has become quite popular in the Tu-160's home country and is often interpreted incorrectly. The advent of the Tu-160 compelled the US government to speed up the development and production entry of the upgraded B-1B Lancer.

Soon afterwards the first photo of the new Soviet bomber (with two Tu-144s standing beside) was circulated in the Western press. Some Western analysts went so far as to suggest that the bomber had been intentionally exposed for the US surveillance satellite for propaganda purposes – like, we got a new bomber! Hear ye! Fear ye! (This explanation appears implausible because *perestroika* and the new Soviet policy of openness were still some years away.) Also, the first available photo was of such appalling quality as to cast doubt on the satellite imagery theory, considering that US surveillance satellites of the early 1980s were reputed to be able to read the number plates of cars! Only later was it revealed that in reality the picture had been taken with a hand-held camera by a quick-minded passenger on a scheduled Aeroflot flight 'terminating at the nearby airport of Bykovo', as press reports claimed. (The latter bit is dubious as well. The approach corridor to Bykovo's runway 12 lies quite a long way north of the LII airfield; during Moscow airshows it could be seen that aircraft on final approach to Bykovo were passing low and far away. At this viewing angle the forest and the hangars would have obscured the Tupolev OKB hardstand from the eyes of passengers on flights inbound to Bykovo. On the other hand, aircraft bound for Moscow-Vnukovo from destinations down south, such as Sochi, often pass directly over the LII airfield at fairly high altitude, making such photos possible; indeed, the angle from which the famous picture was taken indicates a high 'vantage point' – such as a high-flying aircraft.)

The manufacturer's flight tests continued unabated, performed by B. I. Veremey, S. T. Agapov, V. N. Matveyev, V. S. Pavlov and M. M. Kozel. Boris Veremey did more flying on the Tu-160 than anyone else, making 600-plus test flights totalling more than 2,000 hours; as he later put it, he came to feel 'a kinship with this aircraft'. In recognition of his part in 'mastering new military hardware' (a common phrase in those days) Boris I. Veremey was awarded the Hero of the Soviet Union title in 1984. A major contribution to the programme was made at this stage by the then Director of the OKB's flight test facility V. T. Klimov (he went on to become General Director of ANTK Tupolev, as the company is known since 1992), test engineer A. K. Yaschchookov, MMZ No. 156 Chief Engineer A. Mo-zheykov and Kazan' aircraft factory Director V. Kopylov.

The second prototype (*izdeliye* 70-03), likewise unpainted – as it still is – joined the test programme on 6th October 1984, making its first flight that day with OKB test pilot Sergey T. Agapov in the captain's seat. Outwardly it differed from the first prototype in having a recontoured nose radome; the new nose shape was adopted as standard for all subsequent examples. A major milestone was achieved in February 1985 when the first prototype went supersonic for the first time.

(Note: Unlike Western military aircraft (which have *serials* allowing positive identification), since 1955 Soviet/CIS military aircraft normally have two-digit *tactical codes* which are usually simply the aircraft's number in the unit operating it. Three-digit codes are usually worn by development aircraft, often tying in with the construction number (c/n or manufacturer's serial number) or fuselage number (f/n or line number), though some SovAF transports which were previously quasi-civil have tactical codes matching the last three digits of the former civil registration.)

As noted earlier, outwardly the Tu-160 was very similar to the Rockwell B-1, if rather larger. External recognition features included a more pointed nose profile with a sharply raked windshield, the absence of the B-1's characteristic Low-Altitude Ride Control vanes on the sides of the extreme nose (later called Structural Mode Control System), a

Above: This dismembered airframe dumped at the OKB's flight test facility in Zhukovskiy is probably the static test article (*izdeliye* 70-02).

One of the Tu-160 development aircraft (definitely not the first prototype) makes a demo flight at one of the Moscow airshows.

Above: A Tu-160 development aircraft – possibly the second flying prototype (*izdeliye* 70-03) – takes off from runway 30 at Zhukovskiy for a test flight.

The same aircraft after landing on runway 12 at Zhukovskiy; the spoilers are deployed to shorten the landing run. Note the two dielectric panels of different colour built into the fin fillet. The nosecone appears to be all-metal, indicating that no radar is fitted – logical enough for an aerodynamics test vehicle.

Above: The same Tu-160 immediately after touchdown. Note that the air intakes have been modified since the aircraft first flew, featuring six auxiliary blow-in doors instead of the original five; the new design was probably tested on this aircraft before being introduced on production *Blackjacks*.

Another test flight successfully completed, the Tu-160 finishes its landing run. The large structure in the background (of which there are several at Zhukovskiy) serves to shield aircraft not so much from the elements as from snooping satellites.

Above: Tu-160 '29 Grey' takes off for a demonstration flight, trailing the characteristic orange efflux coloured by nitrogen monoxide. Note the single large white dielectric panel on the fin fillet and the three small dark spots at the top of the fin; all the unpainted development aircraft differed in skin shade details.

A Tu-160 makes a high-speed flypast with the wings swept back 65°. The picture clearly shows that the wing fences consist of two segments.

Above: A Tu-160 'cleans up' as it climbs out from Zhukovskiy for a demonstration flight. Note how the flaperons droop in take-off mode.

Despite the 'patchwork' appearance of this particular aircraft, the Tu-160 is an elegant aircraft in any configuration.

Above: This demonstration flight of Tu-160 '29 Grey' ended in an incident – the aircraft lost a good-sized chunk of the fin fillet after a high-speed pass for the crowds. It is seen here landing at Zhukovskiy.

Another view of '29 Grey' from the control tower at Zhukovskiy as it taxies past the watching crowds. The missing dorsal fin section is clearly visible. Note the open APU exhaust port above the nozzles of the port engines.

Above: The first prototype Tu-160 derelict at the OKB's flight test facility at Zhukovskiy in the early 1990s in company with a Tu-95MS and a Tu-22M3; the horizontal tail has been removed.

Another Tu-160 development aircraft parked behind a barbed wire fence at Zhukovskiy. Note the difference in fin treatment.

Above: Tu-160 demonstration flights at Moscow airshows featured spectacular take-offs with a steep banking climbout like this one.

However, attaining high performance is not all about aerodynamics. The Tu-160 had been designed in such a way as to achieve the maximum possible range not only in high-altitude supersonic cruise but also in ultra-low-level terrain-following flight. The bomber's crew was free to choose between these modes or use a combination of them to fulfil the mission with maximum efficiency.

Tu-160 (*izdeliye* 70) production bomber

On 10th October 1984, just four days after the second prototype's first flight, the first Kazan'-built Tu-160 (fuselage number 0101 – ie, Batch 01, 01st aircraft in the batch) took off from the factory's Borisoglebskoye airfield for its maiden flight, captained by Tupolev OKB test pilot Valeriy Pavlov. The second production *Blackjack* (f/n 0102) took to the air on 16th March 1985; the third machine (f/n 0201) followed on 25th December and the fourth (f/n 0202) on 15th August 1986. Production bombers incorporated various aerodynamic refinements, including the abovementioned reprofiled nose, and wore an overall white 'anti-nuclear' colour scheme. (Thus the *Blackjack* is not black at all! Interestingly, the B-1A prototypes were likewise white overall; in contrast, production B-1s are invariably camouflaged.)

Initial-production Tu-160s participated in the trials programme along with the prototypes. The trials did not proceed altogether without incident. On one occasion a Tu-160 captained by Valeriy Pavlov suffered a total electrics failure during a test flight. (The air incident investigation panel later established that the incident had been caused by an overload protection system failure, the possibility of which had not even been considered.) One of the engine-driven generators went to maximum voltage mode, knocking out all the protection systems of the other circuits. The main electric circuits shut down uncommandedly and the few that remained operational worked with excessively high voltage, creating a serious fire hazard. Luckily Pavlov quickly realised what had happened, managing to bring the seemingly hopeless situation under control and bring the bomber home in adverse weather.

Painstaking analysis of the fried electric systems showed that the failure had occurred at the electronic component level and had never been encountered during bench testing. There was no way it could have been discovered, using test equipment and methods then in use. The incident gave rise to a term, 'Pavlov's calculated failure', used for such unforeseeable breakdowns; accordingly the electric systems of all Tu-160s were checked and steps were taken to stop this situation from happening again.

more even upper fuselage contour (the B-1 had a distinctive 'hump' reminiscent of the Boeing 747) and a kinked fin leading edge with a prominent fillet. Also, the nose landing gear unit retracted aft, not forwards (the nosewheel well served for flight deck access), while the main gear units featured distinctive Tu-154 style six-wheel bogies with three pairs of wheels in tandem versus the B-1's four-wheel bogies.

However, the two aircraft had a number of differences which were not so obvious but bore testimony to the different design approach of the Soviet and American engineers. For instance, the original B-1A featured variable supersonic air intakes; at the insistence of the US Air Force these were replaced by simpler fixed-area intakes which restricted the top speed but reduced the aircraft's RCS. Sure enough, the B-1B was formally still a supersonic aircraft which could clock Mach 1.2 at high altitude (versus the B-1A's Mach 2+) but this was not the aircraft's optimum operational mode. Conversely, the Tu-160's variable intakes, coupled with the powerful NK-32 engines and high-fineness-ratio fuselage having a relatively small cross-section area, enabled it to reach 2,200 km/h (1,366 mph; 1,189 kts), as demonstrated in one of the test flights. The rational internal layout of the fuselage (eg, placing the nosewheel well aft of the flight deck rather than below it) also helped reduce drag, as did the carefully optimised forward fuselage contours.

The final assembly shop of the Kazan' aircraft factory (KAPO) in the early 1990s. Two Tu-160s (with f/n 0802 in the foreground) await completion, surrounded by Tu-214 airliners.

Above: The first production Tu-160 built in Kazan' (f/n 0101) pictured during a test flight; the aircraft wore no tactical code. This was the first *Blackjack* to be painted in the overall white colour scheme. Note the five auxiliary blow-in doors and the long pointed fin bullet fairing; both were to change before long.

Below: The first production Tu-160 on the apron of the factory airfield in Kazan'. The aircraft features a production-standard nose profile but the radome is tipped with a pitot boom (compare with the photo on page 35). This aircraft was used for test and development work along with the two Moscow-built prototypes.

Above and below: A spanking new early-production Tu-160 with no tactical code applied yet. Note that the flaperons 'bled down' and the stabilators assumed the maximum nose-up position after engine shutdown.

Above and below: Two more views of the same early-production aircraft. Note the high-gloss finish. The narrow landing gear track enabled the aircraft to use fairly narrow taxiways like this one.

Above: One of the early production Tu-160 'dogships' retained by the Tupolev OKB languishes at the OKB's flight test facility in early September 1993, surrounded by (left to right) a Tu-134AK, a Tu-144D SST and the Tu-134SKh prototype. Note the dirty white colour of the weathered dielectric portions.

The same aircraft two years later in derelict condition (minus engines, stabilisers and a piece of the fin fillet skin) in company with the Tu-134A-3 prototype, two Tu-154B-2s and a Tu-204. The first production Tu-160 is visible in the background (note pitot); sandbags are piled on the cockpit to stop it from falling on its tail.

Launching Tu-160 production at plant No. 22 required new specialised shops to be erected and new technologies mastered. Now the factory possessed unique technological equipment for manufacturing composite and honeycomb-core panels, forging and machining large structural components. The latter included variable-thickness panels with integral stiffeners made of titanium and high-strength aluminium alloys; these components were 20 m (65 ft 7½ in) long, allowing the number of manufacturing joints to be reduced, cutting structural weight and increasing airframe life. The huge wing pivot box carry-through structure was 12.4 m (40 ft 8¼ in) long and 2.1 m (6 ft 10¾ in) wide. It was manufactured in upper and lower halves machined from titanium forgings which were then laser-welded in a vacuum chamber, using special additives and fluxes; this was the Kazan' aircraft factory's proprietary technology. This operation had to be performed strictly at night only, when the city's demand for electricity was lowest, otherwise the Peerless Welding Machine would knock out the electric power in half of Kazan'.

The NK-32 engine officially entered full-scale production in 1986. Previously all Tu-160s had been powered by low-rate initial production (LRIP) engines – actually development engines which had been used in the NK-32's trials programme. Likewise in 1986 the Kh-55SM cruise missile was formally included into the Long-Range Aviation's inventory; in the West the missile was known by its ASCC reporting name AS-15 Kent.

A total of eight aircraft making up the first two production batches (for all practical purposes these were development batches) participated in the Tu-160's trials programme. As already noted, the manufacturer's flight tests and Stage A of the State acceptance trials took place at the Tupolev OKB's Zhukovskiy facility. As the trials progressed, the Red Banner State Research Institute of the Air Force named after Valeriy P. Chkalov (GK NII VVS – *Gosoodarstvennyy krahsnoznamyonnyy naoochno-issledovatel'skiy instiitoot Voyenno-vozdooshnykh sil*) joined in and the action was transferred to the institute's main facility in Akhtoobinsk near Saratov. This location on the Volga River in southern Russia had been chosen with good reason: quite apart from the fact that the steppes of Kazakhstan (used as weapons practice ranges) were conveniently located close at hand, the region had as many as 320 days of fair weather per year.

The team of GK NII VVS pilots conducting the Tu-160's State acceptance trials was headed by L. I. Agoorin. A number of test flights was performed by Lev V. Kozlov who subsequently became the institute's director. Air Force test pilots and navigators Col. M. Pozdnyakov, Col. V. Smirnov, Col. N. Sattarov, Col. S. Popov, Col. V. Neretin, Lt. Col. P. Petrov and Lt. Col. S. Mart'yanov contributed a lot to the *Blackjack*'s trials and eventual service entry.

The rolling steppes beyond the Volga were an ideal proving ground for the Tu-160's main weapon, the Kh-55SM cruise missile with its self-contained navigation system and 3,000-km (1,860-mile) maximum range. During live test launches the bombers were shadowed by an Il'yushin/Beriyev 'Aircraft 976' *Mainstay-C* airborne measuring and control station (AMCS), a purpose-built derivative of the IL-76MD *Candid-B* military transport; five of these radar picket aircraft registered CCCP-76452 through CCCP-76456 were operated by LII. The AMCS monitored the trajectories of the Tu-160 and the missile, using its 360° search radar; it also received telemetry data from the bomber and the missile, tapping it and transmitting it in real time to ground control and telemetry processing centres by radio or satellite link. On several occasions when the Kh-55 started 'acting up' and departed from the designated course, getting too close to the boundaries of the weapons range, a self-destruct command had to be transmitted.

Live weapons trials showed that in maximum range mode the Kh-55SM often pressed on towards the target after the carrier aircraft had landed! In the course of the trials MKB Raduga brought the missile's accuracy margin to a commendable 18-26 m (59-85 ft).

The bomber's mission avionics – the Obzor-K navigation/attack radar and especially the Baikal ESM suite – proved troublesome at first, and a lot of effort was required to bring them up to an acceptable reliability level. The ESM suite was put through its paces at two instrumented test ranges near Orenburg, Russia, and in one of the Central Asian republics. According to press reports (though no documents have been found so far to substantiate this), by mid-1989 the Tu-160 test and development fleet had made a total of about 150 flights; four of them involved Kh-55SM missile launches (on one occasion two missiles were launched simultaneously from the forward and rear bays).

As already mentioned, OKB test pilot Boris I. Veremey attained a top speed of 2,200 km/h (1,366 mph; 1,189 kts) in one of the test flights. In operational service, however, a speed limit of 2,000 km/h (1,240 mph; 1,080 kts) was imposed for structural integrity reasons and in order to conserve service life.

Incidents and accidents of varying seriousness continued as the trials progressed. On one occasion Veremey found himself in a tight spot when the nose gear unit would not extend at the end of a test flight. He opted for a two-point emergency landing and executed it flawlessly, the aircraft suffering only minor

Distinguished Test Pilot Boris Ivanovich Veremey was the Tu-160's project test pilot.

damage and resuming flying after repairs had been made. Another *Blackjack* (the second production aircraft) was not so lucky, crashing immediately after take-off in March 1987; fortunately the entire crew ejected safely.

In addition to long-range test flights and live weapons trials, Air Force test pilots practiced ultra-low-level/terrain-following flight and in-flight refuelling (IFR) techniques, working with Il'yushin IL-78 *Midas* tankers. As the trials went on, the initial-production bombers were prepared, step by step, for delivery to the VVS for evaluation.

On 17th April (some sources say 25th April) 1987 the 184th GvTBAP (*Gvardeyskiy tyazhelobombardeerovochnyy aviapolk* – Guards heavy bomber regiment, = bombardment wing (heavy) in USAF terminology) stationed at Priluki airbase in the Chernigov Region, the Ukraine, took delivery of the first two Tu-160s. As if to underscore the importance of the moment, one of the bombers was ferried by none other than Lt. Gen. Lev V. Kozlov, Deputy Commander of the DA. This was the first time such a complex aircraft was delivered to a first-line service unit for evaluation when the State acceptance trials were still in progress; yet the Minister of Defence Marshal Sokolov had issued appropriate orders which had to be complied with.

As was the case with its US counterpart, the original plans involved a production run of about 100 aircraft. However, the defence spending cuts of the late 1980s brought about by Mikhail S. Gorbachov's *perestroika* changed these plans, and the subsequent collapse of the Soviet Union ruined them completely. By the early 1990s MMZ No. 156 and the Kazan' Aircraft Production Association (KAPO) had built 33 Tu-160s, including static and fatigue test airframes, with two more airframes in various stages of assembly. According to available unclassified sources,

Above: A Tu-160 nudges in to IL-78 CCCP-76675 for an in-flight refuelling. The tanker is trailing a hose and drogue from the fuselage-mounted UPAZ-1 podded hose drum unit.

one *Blackjack* was built in Moscow in 1981, two in 1982 (one in Moscow and one in Kazan', whereupon all production shifted to the latter factory), two aircraft in 1984, one each in 1985 and 1986, four in 1987, five in 1988, three in 1989, five in 1990 and three each in 1991-92. This totals 30 aircraft; the remaining three are unaccounted for.

Speaking of which, Tu-160 batches mostly consisted of five aircraft, as was the case with the Tu-22 *Blinder* and Tu-22M *Backfire* manufactured in Kazan' in earlier days; batches 1 and 2 are an exception, consisting of two aircraft each. As for construction numbers, only one has been identified to date; it is an OKB-owned example coded '63 Grey' (c/n 84704217, which corresponds to f/n 0401). The c/n explanation is probably as follows:

year of manufacture 1988, 4th quarter, *izdeliye* 70, Batch 4, the meaning of the next digit is unknown, 1st aircraft in the batch, completed by Team 7. The c/n is embossed at the top of two large metal plates found on the outer faces of the main gear oleos; these also carry the serial number of the oleo itself but the c/n is the same on both units.

Of this total, 19 production *Blackjacks* were delivered to the 184th GvTBAP, equipping two squadrons; at least three production aircraft were retained by the Tupolev OKB for test and development work, operating from Zhukovskiy. Attrition was limited to the example lost in March 1987.

Considering that so many Tu-160s had remained in the Ukraine after the break-up of the Soviet Union, in the late 1990s Russian President Boris N. Yel'tsin ordered that production of the type be resumed in order to give the Russian Air Force a fully equipped *Blackjack* regiment. In practice this meant completing the unfinished airframes stored at the Kazan' plant. On 10th September 1999 the first new-build Tu-160 in ten years (f/n 0802) made its first flight from KAPO's factory airfield, still in chrome yellow primer finish; on 5th May 2000 it was delivered to the Russian Air Force's 121st *Sevastopol'skiy* GvTBAP as '07 Red'. This was an occasion for much jubilation both in the industry and in the Air Force.

Tu-160PP ECM aircraft project
In the 1980s the Tupolev OKB had a number of upgrades and specialised versions of the Tu-160 on the drawing boards. These included an active ECM variant of the Tu-160 designed to provide ECM cover for large bomber formations. The active jammers and the powerful generators supplying electric power for the mission avionics were to be housed in the weapons bays. The premature termination of the basic bomber's production was the main reason why the Tu-160PP was never built.

Tu-160V bomber project
By far the most radical modification of the *Blackjack* proposed in the 1980s would have resulted in what was, in effect, a completely new aeroplane. Much has been reported on airliner projects utilising cryogenic fuel – liquid hydrogen (LH$_2$) or liquefied natural gas (LNG); the feasibility of the concept has been proved by the Soviet Tu-155 research aircraft – a heavily modified Tu-154 (CCCP-85035) which first flew on 15th April 1988. But have you ever heard of a cryoplane bomber? This is exactly what the projected Tu-160V would have been; the V stood for *vodorod* – hydrogen. Apart from the powerplant consisting of jet engines adapted for running on hydrogen, the Tu-160V featured a new fuselage which was of necessity much bulkier in order to accommodate the heat-insulated LH$_2$ tanks and inert gas pressurisation system.

Other upgrade projects
Among other things, the Tupolev OKB proposed re-engining the Tu-160 with more fuel-efficient Kuznetsov NK-74 afterburning turbofans (which in the event never materialised), installing state-of-the-art mission avionics and integrating new precision weapons. There was also reportedly a projected strategic reconnaissance version.

Tu-160SK suborbital launch vehicle project
Faced by the dwindling defence orders and decline in military hardware production in Russia as a result of political changes and

Wings already at maximum sweep, a Tu-160 breaks formation with an Il-78 tanker.

economic problems, the OKB attempted to find a niche for the Tu-160 on the civil market. In the early 1990s the Tupolev OKB teamed with MKB Raduga and the Moscow Energy Institute (MEI) to develop the Burlak suborbital launch system designed for putting small commercial satellites into orbit. (The name needs to be explained. In 19th-century Russia, the burlaki (pronounced *boorlakee*) were teams of strongmen whose job was to haul barges up rivers by means of ropes.)

The Burlak system comprised the Tu-160SK carrier aircraft (a suitably modified standard *Blackjack*), the Burlak suborbital launcher rocket and a ground support complex, including a launch data preparation system. The Burlak was a three-stage liquid-propellant rocket with a launch weight of 20 tons (44,090 lb); it was to be suspended under the centre fuselage on a special pylon. The maximum payload put into orbit by the rocket was 800-850 kg (1,760-1,870 lb) and the calculated launch costs amounted to US$ 6,000-8,000 per kilogram of satellite weight.

Since the Tu-160SK retained the baseline bomber's IFR system, it could launch the Burlak almost anywhere, depending on where the launch conditions were favourable at the moment. Estimated unrefuelled range with the launcher rocket carried internally instead of externally was 1,000 km (620 miles). Launch was to take place at 9,000-14,000 m (29,530-45,930 ft) with the aircraft cruising at 850-1,600 km/h (530-990 mph; 460-865 kts). If necessary the Tu-160SK carrying a refuelled but unequipped Burlak could land at a predesignated military base in any country so that the satellite could be mated with the launcher rocket. In so doing due precautions would be taken to guarantee protection of the customer's technology against industrial espionage. All the satellite's manufacturer had to do was make sure that certain design parameters (electric system, dimensions, weight, operational temperatures etc.) met the requirements of MKB Raduga so that the satellite would be compatible with the rocket.

The Burlak system (also called Burlak-Diana) permitted the launch of several satellites at a time, providing their aggregate weight and dimensions were within the specified limits. Preliminary calculations showed that a high-speed, high-altitude suborbital launch required two to three times less energy to place an equivalent payload in orbit as compared to a conventional launch. Also, the costs would be 2-2.5 times lower.

Above: Tu-160 '86 Grey' in low-speed flight with the flaps, slats and flaperon at maximum deflection. Note the abbreviated fin bullet fairing introduced to reduce vibrations which ruined the defensive avionics.

The ability to launch the rocket over the World Ocean allowed satellites to be placed into any orbit while obviating the need to lease land for restricted areas where launcher rocket stages would drop after burnout – an issue which could lead to international disputes. The performance parameters of the Burlak system were far superior to those of its American counterpart – the Orbital Systems Corporation-Hercules Pegasus suborbital launcher rocket which was carried aloft by a subsonic Boeing B-52 Stratofortress bomber. Thus, the Tu-160 strategic bomber offered unparalleled opportunities for commercial use.

Russia endeavoured to turn the Burlak-Diana programme into an international one, searching for foreign partners. Thus on 2nd

Close-up of the forward and centre fuselage of '86 Grey', showing the original auxiliary blow-in door arrangement, the flap sealing plates and the black radar absorbing material (RAM) coating of the air intakes. The tactical code is carried only on the port nosewheel well door.

Above: Tu-160 c/n 84704217 at the MAKS-95 airshow, mated with a full-scale mock-up of the Burlak suborbital launcher rocket to represent the Tu-160SK.

Another view of the same aircraft, showing the exhibit code '342' with which it had been displayed at the 1995 Paris Air Show. Note the extreme deflection of the fin and stabilisers.

Still in primer finish, the latest *Blackjack* delivered to date (f/n 0802) is pictured during its first flight on 10th September 1999. It was delivered to the 121st GvTBAP at Engels-2 AB on 5th May 2000 to become '07 Red' *Aleksandr Molodchiy*.

April 1997 the Russian government issued directive No. 428 ordering the Ministry of Economics, the Ministry of Foreign Trade (jointly with the Ministry of Defence and the Ministry of Foreign Affairs), MKB Raduga, ANTK Tupolev and the Rosvo'oroozheniye State Company (which was then the nation's main arms exporter) to hold negotiations with Deutsche Aerospace (the German space agency) and the company OHB-System GmbH on possible technical partnership in the programme. However, no tangible results were forthcoming as of this writing.

MKB Raduga, the OKB of the Moscow Energy Institute, ANTK Tupolev and OHB-System GmbH unveiled the Tu-160SK and the Burlak-Diana two-stage rocket in model form at the Asian Aerospace '94 airshow which took place at Singapore-Changi airport. In July 1995 a real *Blackjack* with no tactical code (ex '63 Grey', c/n 84704217) was displayed at the 41st Paris Air Show with a full-scale wooden mock-up of the Burlak-Diana rocket, bearing the exhibit code 342. The same aircraft with the same mock-up was also in the static park of the MAKS-95 airshow in Zhukovskiy (22nd-27th August); it was present again at the MAKS-97 airshow (19th-24th August 1997).

Interest in the Tu-160/suborbital launch idea was unexpectedly revived when the Ukraine started disposing of its *Blackjack* fleet in 1999. A US company called Platforms International Corporation (PIC) and based in Mojave, California, offered to buy three of the Ukrainian aircraft, plus spares, for US$ 20 million and transfer 20% of the shares of its division, Orbital Network Services Corporation (OrbNet), to the Russian Aerospace Consortium. The Ukrainian Cabinet of Ministers approved the sale, and it seemed that OrbNet would be able to launch the first satellite in a year of two. Still, this project never materialised.

Tu-160 Voron strategic reconnaissance drone carrier project

In 1968 or 1969 the Soviet Union laid hands on a most unusual piece of American aviation hardware – one of the 38 Lockheed D-21 (GTD-21B) supersonic reconnaissance drones. This aerial vehicle shared certain design features with the famous Lockheed SR-71 Blackbird Mach 3 reconnaissance aircraft; in fact, it looked like a scaled-down SR-71 engine nacelle with wings. Of course the captured D-21 was subjected to detailed analysis with the participation of the leading Soviet aircraft, electronics and defence industry enterprises – including the Tupolev OKB. On 19th March 1971 the Military Industrial Commission of the USSR Supreme Soviet issued Ruling No. 57, prescribing the Tupolev OKB to develop a Soviet analogue of the D-21, making use of indigenous structural materials, engines and equipment. The projected reconnaissance drone was provisionally designated **Vor**on (Raven) – no doubt because the D-21 was flat black overall, being covered with a special heat-dissipating paint.

As one might imagine, the Voron (no numeric OKB designation is known) was extremely similar in appearance to the D-21 – right down to the two pitot booms flanking the air intake with its shock cone. The main external difference lay in the shape of the wings which were close to a pure delta planform, whereas those of the D-21 had large curved LERXes similar in shape to the SR-71's nose chines. As was the case with the American drone, the cameras and their film cassettes were to be housed in a special capsule which was ejected and retrieved after the drone had passed over the target. (Incidentally, this technology was nothing new to the Tupolev OKB; on the Tu-123 Yastreb supersonic reconnaissance drone built in quantity for the Soviet Air Force the entire forward fuselage housing the cameras was jettisoned and parachuted to safety over territory held by friendly troops.)

The Voron was to be powered by a 1,350-kgp (2,975-lb st) RD-012 supersonic ramjet

55

and accelerated to ramjet ignition speed by a massive solid-propellant rocket booster attached to the underside and delivering an awesome 47,500 kgp (104,720 lb st). Dry weight was estimated as 3,450 kg (7,605 lb) and own launch weight as 6,300 kg (13,890 lb), increased to 14,120 kg (31,130 lb) by the rocket booster. Overall length was 13.06 m (42 ft 10 in), wing span was 5.8 m (19 ft ¼ in) and height was 2.08 m (6 ft 10 in); wing area was 37.0 m² (397.85 sq ft). Design cruising speed was 3,500-3,800 km/h (2,170-2,360 mph; 1,890-2,055 kts), operational altitude was 23,000-26,400 m (75,460-86,615 ft) and maximum range was 4,600 km (2,855 miles).

Now where does the Tu-160 come in? Quite simply, the Voron was to be launched by suitably modified Tu-160 or Tu-95 bombers – just like the D-21 was launched by modified B-52s; ground launch was deemed to be less effective. Hence the drone launcher version of the *Blackjack* was provisionally known as the Tu-160 Voron. Eventually, however, the bird never hatched and the project died with it.

Above left: 184th GvTBAP Tu-160 '16 Red' was shown to French Defence Minister Jean-Pierre Chèvenement at Kubinka AB in March 1993.

Left: A Tu-160 in late-production configuration cruises with the wings at 35°.

Below: '63 Grey' seen on short finals to Zhukovskiy is interesting in combining the late-model six auxiliary blow-in doors with the original long bullet fairing.

Chapter 4

The Tu-160 in Detail

The Tu-160 is a multi-mode strategic bomber and missile strike aircraft designed to deliver weapons from both high and low altitude. It utilises a blended wing/body (BWB) layout which maximises the use of internal space for weapons, fuel, avionics and equipment while reducing the number of manufacturing joints, thereby cutting structural weight.

The airframe is of basically all-metal construction, being made largely of high-strength aluminium alloy; titanium and steel are used for critical components absorbing the main structural loads. Some non-stressed elements (such as wing/fuselage fairings and weapons bay doors) and the trailing edge portions of the tail surfaces are made of glass-fibre/textolite composite, in some cases with a honeycomb core. The airframe subassemblies are joined together by bolts and rivets. Numerous access hatches and removable panels are provided for maintenance purposes.

Fuselage: Semi-monocoque stressed-skin structure made up of four sections (No. 1 and No. 2 forward, centre and aft fuselage). The skin is reinforced with frames and stringers; cutouts for hatches are reinforced by longitudinal beams.

The *No. 1 forward fuselage section (Section F-1)* comprises the nosecone, the flight deck and the nosewheel well. The lower half of the nosecone is a radar antenna bay enclosed by a dielectric radome; the upper half incorporates a fully retractable refuelling probe. The pressurised flight deck seats four, the captain and co-pilot sitting up front with the navigator and weapons systems operator behind them; it is accessed via a pressure door in the nosewheel well, using a mobile ladder. The flight deck windscreen consists of an optically-flat Triplex centre panel and two curved Plexiglas side quadrants; there are also four small side windows, the forward two of which are sliding direct vision windows, and two eyebrow windows; a single side window is provided on each side for the navigator and WSO. A roof section over each crew member's seat is jettisoned in the event of ejection; it can also be removed on the ground for ejection seat maintenance.

Section F-1 also incorporates avionics bays. It is of quasi-oval cross-section.

The *No. 2 forward fuselage section (Section F-2)* has an integral (BWB) layout, being manufactured as one with the fixed inner wing sections. The leading edge root extensions incorporate integral fuel tanks, while the

The forward fuselage of Tu-160 '05 Red' (one of two to be named *Il'ya Muromets*). Note the rear-vision periscope over the WSO's window, the faired window of the OPB-15T bomb sight, the dorsal and ventral communications blade aerials and the red covers over angle of attack vanes and static ports.

57

fuselage proper incorporates the forward weapons bay. It also mounts the wing pivot box carry-through structure to which the outer wings are attached.

The *centre fuselage (Section F-3)* incorporates the front portion of the aft weapons bay (which is flanked by the mainwheel wells located between fuselage frames 54 and 65), avionics bays and the APU bay. It also carries the engine nacelles.

The *aft fuselage (Section F-4)* incorporates the rear portion of the aft weapons bay, equipment bays and three integral fuel tanks. It also serves as an attachment point for the fixed lower section of the fin. The rearmost portion houses the brake parachutes and the defensive avionics (ESM) suite.

Wings: Cantilever low-wing monoplane with variable sweepback. The centre section is integral with the fuselage. The trapezoidal outer wings are five-spar structures with chemically milled skins and only six ribs each; they are attached via pivot bearings to a massive wing pivot box carry-through unit made of titanium. Wing sweep is varied between three settings (20°, 35° and 65°) by means of hydraulic actuator rams attached to the front spar of the wing centre section.

The outer wing torsion boxes serve as integral fuel tanks. The wing leading and trailing edges house control runs, control surface actuators, flap and slat drive shafts and electric cables.

To improve low-speed handling and lift/drag ratio the outer wings are equipped with four-section leading-edge slats and three-section double-slotted flaps; the latter feature movable curtains closing the gap between the rear spar and the foremost flap segment. The outer wings are also equipped with six-section spoilers/lift dumpers and flaperons.

A curious feature of the Tu-160's wing design is that the innermost portions of the outer wing trailing edges, together with the triangular segments of the inboard flap sections, hinge upwards as wing sweep is increased to form a sort of boundary layer fence at maximum sweep. This feature opti-

Top: The pilots' instrument panels and central console. The captain sits on the right, working the bank of throttles offset to the right. Note the pedestal-mounted control sticks; the landing gear control lever is on the captain's instrument panel. The twin red handles just visible in the lower right-hand corner are the actuation handles of the captain's K-36DL ejection seat.

Bottom: A view of the pilots' workstations in the Tu-160. Although the instruments are predominantly of the analogue type, there are two large multi-function displays at the top of the panel. Note the plethora of buttons and switches, the rubber-bladed ventilation fans characteristic of Soviet aircraft and the red release handles of the flight deck roof escape hatches.

mises the flow around the inner/outer wing joint at high speed while obviating the need to have the outer wing trailing edges slide inside the wing gloves, an area which is quite tricky to seal in order to minimise drag.

Wing span is 57.7 m (189 ft 3¾ in) at minimum sweep (20°), 50.7 m (166 ft 4 in) at intermediate sweep (35°) and 35.6 m (116 ft 9½ in) at maximum sweep (65°). Wing aspect ratio at these settings is 6.78, 5.64 and 2.85 respectively. Total wing area 293.15 m² (3,152.15 sq ft), including 189.83 m² (2,041.18 sq ft) for the outer wings; flap area 39.6 m² (425.8 sq ft), leading-edge slat area 22.16 m² (238.28 sq ft), flaperon area 9 m² (96.77 sq ft) and spoiler area 11.76 m² (126.45 sq ft).

Tail unit: Cruciform swept tail surfaces. The all-movable stabilisers of trapezoidal planform are rigidly connected by a centre section member into a single unit. Structural strength problems encountered soon after service entry necessitated a reduction of the horizontal tail span to 13.26 m (43 ft 6 in); leading edge sweep is 44°, aspect ratio 3.16 and overall area 55.6 m² (597.85 sq ft).

The fin is made up of two sections, the lower one extending forward into a large fin root fillet incorporating dielectric panels; the trailing edge of the fin lower section carries ESM antennas. The cropped-delta upper fin section is all-movable, serving for directional control. The fin is 6.95 m (22 ft 9½ in) tall; leading edge sweep is 47°, aspect ratio 1.15 and overall area 42.025 m² (451.88 sq ft), including 19.398 m² (208.58 sq ft) for the upper section. There is a large bullet fairing extending aft at the fin/stabiliser junction.

Landing gear: Hydraulically-retractable tricycle type. Wheel track 5.5 m (18 ft ½ in), wheelbase 17.8 m (58 ft 4¾ in). The steerable nose unit retracts aft; it is fitted with twin 1,080 x 400 mm (42.5 x 15.7 in) wheels and an electrohydraulic steering actuator controlled by the rudder pedals. The nosewheels are equipped with a mud/snow/slush guard to prevent foreign objects from being thrown up into the engine air intakes.

Top: Close-up of the flight deck glazing. The foremost side window on each side is a sliding direct-vision window which can be used for emergency evacuation on the ground.

Centre: A curious feature of the Tu-160's wing design is that the innermost portions of the outer wing trailing edge and triangular inboard sections of the wing flaps fold upwards as wing sweep increases to form boundary layer fences.

Bottom: The Tu-160 has all-movable tail surfaces. The conduits running along the fuselage sides (with heat shields) are for the rear fuselage fuel tanks. Note the ESM antennas on the fin trailing edge and the canvas covers on the engine nozzles.

Left: The Tu-160's main landing gear units have six-wheel bogies to reduce runway loading. The many holes in the wheel hubs admit air for better brake cooling.

Centre: The main gear units feature twin transverse hinges which let the bogies move inwards during retraction to lie in bays inboard of the engine nacelles while still providing the widest possible wheel track. Note the sloping retraction struts.

Bottom: The nose gear unit of '05 Red'.

The semi-levered suspension main units retract aft into the centre fuselage and have telescopic retraction struts attached at the front. They are fitted with six-wheel bogies (three pairs of wheels in tandem) with 1,260 x 425 mm (49.6 x 16.7 in) wheels and multi-disc brakes. During retraction the bogies rotate 180° aft to lie inverted in the wheel wells. Additionally, the main units feature double transverse hinges which allow them to move inwards and shorten during retraction, reducing the landing gear track by 1.2 m (3 ft 11¼ in) at the same time. Soon after service entry the design of the main gear units was simplified to improve reliability.

All three units feature oleo-pneumatic shock absorbers; those of the main units are of three-chamber design. The nosewheel well is closed by twin lateral doors; each mainwheel well is closed by twin lateral main doors and an auxiliary door segment attached to the retraction strut. All doors remain open when the gear is down.

Powerplant: Four Kuznetsov NK-32 afterburning turbofans rated at 13,000 kgp (28,660 lb st) dry and 25,000 kgp (55,115 lb st) reheat. The NK-32 is a three-spool turbofan with a three-stage low-pressure (LP) compressor, a five-stage intermediate-pressure (IP) compressor, a seven-stage high-pressure (HP) compressor, a multi-burner annular combustion chamber, single-stage HP and IP turbines, a two-stage LP turbine, a core/bypass flow mixer, an afterburner and a convergent-divergent axisymmetrical nozzle.

The air intake assembly has a fixed spinner and 18 inlet guide vanes (IGVs). The HP compressor incorporates bleed valves supplying air for the de-icing system, pressurisation and air conditioning system etc. Two of the combustion chamber's flame tubes feature igniters. The HP turbine has monocrystalline blades.

A ventral accessory gearbox is mounted near the front end of the engine; it features an integral constant-speed drive (CSD) for the AC and DC generators and hydraulic pump. The NK-32 has a self-contained lubrication system, a full authority digital engine control system (FADEC) and test equipment monitoring powerplant operation. The engine is started

Right: Another view of the nose landing gear unit, showing the collapsible drag strut, the steering mechanism/shimmy damper cylinders and the massive mud/snow/slush guard. Note the mobile stairs serving for access to the flight deck via the nosewheel well.

Bottom: The air intakes of the port engines. This view shows the black radiation-absorbing material coating of the inlet ducts, the intake upper lip acting as a boundary layer splitter plate and the red covers on the sharp lower lip to protect the ground personnel against injury.

by an air turbine starter (ATS) using compressed air supplied by the APU or a ground source.

The NK-32 is 7.453 m (24 ft 5½ in) long, with a maximum diameter of 1.7 m (5 ft 7 in); dry weight is 3,650 kg (8,046 lb). The bypass ratio is 1.36, the overall engine pressure ratio 28.2 and the turbine temperature is 1,630° K.

The engines are housed in two paired nacelles adhering directly to the wing centre section underside; each engine is attached to the nacelle structure by struts via three mounting rings built into the outer casing. Each nacelle features a bifurcated supersonic intake (V-shaped in plan view) with multi-section vertical flow control ramps; the upper lip of the intake stands apart from the wing undersurface, acting as a fixed boundary layer splitter plate. The inlet duct cross-section changes gradually from rectangular at the intakes to circular at the compressor faces. The forward section of each inlet duct incorporates six spring-loaded auxiliary blow-in doors located on the outer and inner faces of the nacelle; these open to supply additional air at high rpm.

An Aerosila (Stoopino Machinery Design Bureau) TA-12 auxiliary power unit (APU) supplies compressed air for engine starting and air conditioning, as well as ground/emergency AC/DC power for systems and equipment (TA = *toorboagregaht* – lit. 'turbo unit'). The APU is located in an unpressurised bay in the centre fuselage.

Control system: The Tu-160 was the first Soviet series-produced heavy aircraft to feature a triply redundant fly-by-wire (FBW) control system. Should all four FBW control channels fail, there is an emergency mechanical flight control system as a last resort. This makes for high control reliability in all flight modes.

The control system consists of various mechanical, hydromechanical, electrohydraulic, electromechanical, electronic and electric devices. It comprises the control surface actuation system, the automatic flight control system (AFCS – ie, autopilot and automatic approach/landing system) and the wing control system. The aircraft has full dual controls; unlike most heavy bombers, the flight deck features fighter-type control sticks

The engine nacelles of one of the prototypes (above) and an early-production Tu-160 (below). Note the black RAM coating on the latter aircraft.

instead of the usual control wheels. All control surfaces are powered by irreversible hydraulic actuators; there is an artificial-feel mechanism. The AFCS receives inputs from the control sticks and rudder pedals, the system's own sensors and the sensors and computers of other systems.

The wing control system optimises the wings' configuration to the respective flight modes. It comprises the high-lift device control system, the wing sweep control system and the wing fence actuation system.

Fuel system: 13 integral fuel tanks are located in the inner wings/fuselage and the outer wing torsion boxes; the total fuel load is 171 tons (376,980 lb). (Note: Some sources quote a fuel load of 148 tons (326,280 lb).) The fuel tanks are split into four groups, one for each engine, with a cross-feed system. Each group has a service tank from which fuel is fed to the respective engine; to this tank the fuel is delivered by electric fuel transfer pumps. Fuel is also transferred between forward and aft tanks in flight to maintain CG position.

A fully retractable in-flight refuelling probe with a GPT-2 connector (*golovka toplivopreeyomnika*) is installed ahead of the windshield, enabling the Tu-160 to take on fuel from IL-78/IL-78M *Midas-A/B* tankers. A fuel jettison system is provided to reduce the landing weight in the event of a forced landing.

Hydraulics: Four separate hydraulic systems, each powered by a single pump driven via the respective engine's accessory gearbox. Auxiliary turbopumps are provided to supply hydraulic power on the ground or in an emergency (in the event of a quadruple engine failure).

The hydraulic systems power the control surface actuators, air intake ramps, high-lift devices, wing actuators, landing gear and the rotary launchers in the weapons bays. The systems and their units are remote-controlled ('power-by-wire'); nominal pressure is 280 kg/cm^2 (4,000 psi).

Electrics: AC power supplied by four engine-driven generators with integral CSDs and a fifth generator driven by the APU. DC power supplied by four engine-driven brushless generators; backup DC power provided by batteries. The electric system includes power distribution panels and circuit breakers. A ground power receptacle is provided.

De-icing system: The wing leading edge and engine inlet guide vanes are de-iced by engine bleed air. Electric de-icing on the fin and stabiliser leading edges, engine air intakes, windshield, pitot heads, and static ports. There is a radioactive isotope icing sensor.

This view shows how the flattened underside of each engine nacelle gradually changes to the cylindrical cowlings.

Air conditioning and pressurisation system: The ventilation-type flight deck and avionics bays are pressurised by engine bleed air to ensure comfortable working conditions for the crew and normal operating conditions for the avionics throughout the flight envelope. Pressurisation air is cooled by heat exchangers and filtered before being fed to the flight deck and avionics bays at reduced pressure.

Oxygen system and crew gear: All crew members are provided with oxygen masks, standard Air Force G-suits and ZSh-7B protective helmets (*zaschchitnyy shlem*) with tinted visors. For extreme high-altitude missions the crew wears Baklan (Cormorant) pressure suits featuring full-face pressure helmets similar to those worn by astronauts.

Fire suppression system: A centralised fire suppression system with flame sensors is provided for extinguishing fires in the engine nacelles and the APU bay. Hand-held fire extinguishers are provided in the flight deck.

Brake parachute system: Three cruciform main brake parachutes of 35 m^2 (376.34 sq ft) each are deployed by two 1m^2 (10.75 sq ft) drogue parachutes to shorten the landing run or in the event of an aborted take-off. The brake parachute container is located ventrally in the aft fuselage between frames 98 and 100; the parachutes are deployed electro-pneumatically.

Avionics and equipment: The Tu-160 is equipped with state-of-the-art flight and mission avionics, including a specially-developed weapons control system. The avionics enable automatic flight along a predesignated route and accurate delivery of all weapons the aircraft is compatible with, day and night, in any weather. The avionics suite features more than 100 digital computers.

a) navigation and piloting equipment: Duplex K-042K inertial navigation system (INS), stellar navigation system and satellite navigation equipment. Sopka (Hill) ground avoidance radar in nose radome for ultra-low-level terrain-following flight; see also targeting equipment.

b) communications equipment: Multi-channel digital communications suite, including an intercom for crew communication. Communications blade aerials are located above and below the flight deck.

The upper rear sections of the engine nacelles are skinned in heat-resistant steel. This photo also shows the petals of the No. 4 engine's variable nozzle.

Left: The Tu-160 features a fully retractable in-flight refuelling probe enabling the bomber to take on fuel from IL-78 tankers. Unfortunately, lack of adequate funding of the Russian Air Force (and hence lack of adequate proficiency training) prevents the *Blackjack* crews from practising IFR procedures as often as they would like to!

Below: The underside of the Tu-160's centre fuselage is occupied by two long weapons bays separated by the wing pivot box carry-through structure. Compare this to the B-1 which has three shorter weapons bays (see page 34).

Opposite page: Each of the Tu-160's weapons bays houses an MKU-6-5U rotary launcher. The lower photo shows a Kh-55SM missile in long-range configuration with conformal tanks suspended on the aft rotary launcher; note also the hydraulic system components on the weapons bay walls.

c) targeting equipment: Obzor-K navigation/attack radar in nose radome for target illumination and ground mapping; it is capable of detecting ground and sea targets (surface ships) at long range and providing target data to semi-active radar homing (SARH) air-to-surface missiles. OPB-15T electro-optical bomb sight (op***tich***eskiy pri***tsel*** bombardi***rov***ochnyy, televizi***on***nyy) installed in a teardrop fairing with an optically-flat window under the flight deck for bomb-aiming in daylight/low light level conditions. Provisions for laser designator allowing laser-guided bombs to be integrated. Sproot-SM (aka SURO-70) system for preparing and downloading target data to Kh-55SM cruise missiles (SURO = sis***tema*** oopravl***en***iya ra***ket***nym o***roo***zhiyem – missile control system).

d) identification friend-or-foe (IFF) system: SRO-1P *Parol'* (Password, aka *izdeliye* 62-01) IFF transponder (samolyotnyy rahdi-olokatseeonnyy otvetchik – lit. aircraft-mounted radar responder). The triangular IFF aerials are located under the nose (aft of the radome) and on top of the tailcone.

e) electronic support measures (ESM) equipment: The Baikal ESM suite includes a radar homing and warning system (RHAWS) with aerials on the forward/aft fuselage sides and wingtips to give 360° coverage. Active ECM equipment with antennas under flush dielectric panels in the LERXes and small rounded fairings on the lower fin section trailing edge. Active infra-red countermeasures (IRCM) equipment consists of an IR missile warning system covering the rear hemisphere. Passive ECM and IRCM provided by numerous three-round APP-50 chaff/flare dispensers built into the aft fuselage underside pivot joints; white rear navigation light on tail-cone. Two retractable landing lights just aft of the radome, augmented by two landing/taxi lights on the nose gear oleo. White anti-collision strobe lights on top of centre fuselage and under the starboard engine nacelle. Refuelling probe illumination light buried in the fin leading edge. Pilot lights on all three landing gear struts to show the ground personnel that the gear is down and locked during night approaches.

Crew rescue system: The crew rescue system comprises the access hatch jettison system, four Zvezda K-36DL zero-zero ejection seats, an ejection sequencing system and NAZ-7 survival kits (*nosimyy ava**ree**ynyy za**pahs***). An LAS-5M inflatable rescue dinghy is provided in case of ditching or overwater ejection.

The crew members may eject individually, or any crew member may initiate ejection, punching everybody out if any of the crew are disabled.

Armament: The Tu-160's maximum ordnance load is 40 tons (88,180 lb). All armament is carried internally in two tandem weapons bays measuring 11.28 x 1.92 m (37 ft x 6 ft 3½ in).

The principal armament consists of air-to-surface missiles designed for delivering nuclear strikes against targets with known co-ordinates. The standard ASM type carried by the Tu-160 is the Raduga Kh-55SM (AS-15B *Kent*) strategic air-launched cruise missile; 12 of these weapons can be carried with wings and tail surfaces folded and the engine stowed (six in each weapons bay on an MKU-6-5U rotary launcher). The Kh-55SM is 8.09 m (26 ft 6½ in) long and has a launch weight of 1,700 kg (3,750 lb); maximum range is 3,000 km (1,860 miles), thanks to the optional conformal fuel tanks on the fuselage sides. Before launch a digital map of the route to the target is fed into the missile's computer by the SURO-70 MCS. Then the missile is pneumatically ejected from the launcher; seconds later the wings and tail unit unfold, the turbofan engine is deployed and started up.

The *Blackjack* also has provisions for carrying the Raduga Kh-15S (AS-16 *Kickback*) designed for destroying the enemy's air defence system. 24 of these short-range ASMs can be carried (12 on each rotary launcher); in this case the Tu-160 operates in low-level mode.

Additionally, from the start the Tu-160 was intended to carry other types of weapons, including conventional free-fall bombs, hence the provision of the OPB-15T precision electro-optical bomb sight.

A Kh-55SM cruise missile (long-range version) in flight configuration on a ground handling dolly. Note how the aft portions of the conformal tanks are cut away from above to clear the wings.

to fire 50-mm (1.96-in) magnesium flares or chaff bundles as a protection against air-to-air and surface-to-air missiles.

f) flight instrumentation: Similar to that of the Tu-22M3 *Backfire-C*, with largely electro-mechanical instruments but also several multi-function displays (MFDs).

g) data recording equipment: Standard MSRP-64-2 flight data recorder (FDR) and Mars-BM cockpit voice recorder (CVR) for mission debriefing or accident analysis.

h) exterior lighting: Port (red) and starboard (green) navigation lights in wingtips and on the inner wing leading edges near the

Tu-160 specifications

Powerplant	4 x Kuznetsov NK-32
Thrust, dry/reheat, kgp (lb st)	4 x 13,000/25,000
	(4 x 28,660/55,115)
Length overall	54.095 m (177 ft 5¾ in)
Wing span (at 20°/35°/65° sweep)	57.7/50.7/35.6 m
	(189 ft 3¾ in/166 ft 4 in/116 ft 9½ in)
Height on ground	13.0 m (42 ft 8 in)
Total wing area, m² (sq ft)	293.15 (3,152.15)
Movable outer area, m² (sq ft)	189.83 (2,041.18)
Leading-edge sweep	20°/35°/65°
Wing aspect ratio at 20°/35°/65° sweep	6.78/5.64/2.85
Landing gear track	5.5 m (18 ft ½ in)
Landing gear wheelbase	17.8 m (58 ft 4¾ in)
Maximum take-off weight, kg (lb)	275,000 (606,260)
Maximum landing weight, kg (lb)	155,000 (341,710)
Maximum ordnance load, kg (lb)	40,000 (88,180)
Maximum speed, km/h (mph; kts)	2,000 (1,240; 1,080)
Landing speed at 140,000-150,000 kg (308,640-330,690 lb) LW, km/h (mph; kts)	260-280 (161-174; 140-151)
Rate of climb, m/sec (ft/min)	60-70 (11,800-17,710)
Service ceiling, m (ft)	15,000 (49,210)
Range in supersonic cruise, loaded, km (miles)	2,000 (1,240)
Take-off run at 150,000-275,000 kg (330,690-606,260 lb) TOW, m (ft)	900-2,200 (2,950-7,220)
Landing run at 140,000-155,000 kg (308,640-341,710 lb) landing weight, m (ft)	1,200-1,600 (3,940-5,250)
G limit	2

Chapter 5

In Soviet Service

And in Later Days

As recounted in Chapter 3, Distinguished Military Pilot Maj. Gen. Lev V. Kozlov – the then Deputy Commander (Combat Training) of the Soviet Air Force's long-range bomber arm – was the first service pilot to fly the Tu-160. (Distinguished Military Pilot is an official grade reflecting experience and expertise.) His next-in-command, 37th VA Commander Pyotr S. Deynekin (then in the rank of Lieutenant-General), was just as quick to master the new type. We'll let the Tu-160's project test pilot Boris I. Veremey tell the story:

'Lev Vasil'yevich Kozlov would come from the Long-Range Aviation headquarters in his car to pick me up and we would go to Zhukovskiy to perform a standard [familiarisation training] programme which included between seven and fourteen flights. I was also the instructor pilot when Pyotr Stepanovich Deynekin took his training [on the Tu-160]. He made seven flights with high class, flying the aircraft smoothly and beautifully. Marshal Yefimov, [the then] Commander-in-Chief of the Air Force, had banned the Long-Range Aviation Commander from flying, citing safety concerns as the reason. Thus Deynekin had to make his flights clandestinely while the C-in-C was on vacation.'

In April 1987 the Tu-160 achieved initial operational capability (IOC) with the 184th *Poltavsko-Berlinskiy* Red Banner GvTBAP based in the Ukrainian town of Priluki. This unit had fought with distinction during the Great Patriotic War, earning the prestigious Guards status for gallantry above and beyond the call of duty; it was awarded the Order of the Red Banner of Combat (one of the highest military decorations in the Soviet Union) and earned the honorary titles *Poltavsko-Berlinskiy* for its part in liberating the Ukrainian city of Poltava and in taking Berlin.

After the war the 184th GvTBAP remained one of the Soviet Air Force's top-notch units. It had been the first to convert to the Tu-4 *Bull* long-range bomber; later the regiment operated various versions of the Tu-16 *Badger* medium bomber, and in 1984 it re-equipped with the then-latest Tu-22M3. That said, it was no coincidence that the unit was chosen to introduce the *Blackjack* into service. The advent of the Tu-160 brought about a major reconstruction of the Priluki airbase; among other things the runway was reinforced and extended to 3,000 m (9,840 ft).

The pilots and technicians of the 184th GvTBAP had to master the Tu-160 while the State acceptance trials were still in progress. This was because the trials looked set to be a protracted affair due to the large scope of the work necessitating a large number of test flights. The decision to start Tu-160 operations (or, to be precise, evaluation of the new bomber) made it possible to assess the *Blackjack*'s strengths and weaknesses, ironing out bugs and accumulating first-hand experience which other units slated to re-equip with the type would find invaluable.

Operating such a complex aircraft, especially during the service introduction phase,

An early-production Tu-160 in a revetment at Priluki AB, with a jet blast deflector in the background. Note the tall stepladder which is part of the Tu-22M's ground support equipment and is used for accessing the *Backfire*'s cockpits which have individual upward-opening canopies.

Above: A Tu-160 at a display of the latest Soviet military hardware at Kubinka AB in 1989 organised for the top Soviet government officials; note the Beriyev A-40 Albatross ASW amphibian in the background.

demanded higher-than-average skill and responsibility on the part of the flight and ground crews. Usually Long-Range Aviation personnel took conversion training on new types at the DA's 43rd Combat & Conversion Training Centre (TsBP i PLS – *Tsentr boyevoy podgotovki i pereoochivaniya lyotnovo sostahva*) at Dyaghilevo AB near Ryazan'. This time, however, the training took place right at the factories; the flight crews were trained at aircraft factory No. 22 in Kazan', while the technicians studied the NK-32 at engine factory No. 24 in Kuibyshev. A priority task for Tupolev OKB test pilot Boris I. Veremey and the factory test pilots in Kazan' was to train qualified flying instructors (QFIs) for the 37th VA of which the 184th GvTBAP was part; the QFIs would then pass on their skills to service pilots transitioning to the *Blackjack*.

Not that training the flight crews was excessively complicated; it was done using Tu-134UBL *Crusty-Bs*. The purpose-built Tu-134UBL (*oochebno-boyevoy* [*samolyot*] *dlya lyotchikov* – combat trainer for pilots) is a derivative of the Tu-134B *Crusty-A* 80-seat short/medium-haul airliner, designed for training heavy bomber crews. Its *raison d'être* was that, unlike the earlier Tu-22 *Blinder*, the Tu-22M and Tu-160 did not have specialised trainer versions, and using the bombers for conversion/proficiency training would be a waste of their service life. Also, the Tu-134 is similar to the *Backfire* and *Blackjack* in thrust/weight ratio and low-speed handling.

Tu-22M pilots found the Tu-160 to be much easier to fly. One can only imagine how tiring the *Backfire* was to fly, considering that a *Blackjack* pilot has to work all manner of switches like a pianist works the piano keys, performing more than 2,000 switch operations within 15 minutes. On final approach, for instance, a Tu-160 captain has to perform *two operations per second*!

As already mentioned, the first two *Blackjacks* destined for the 184th GvTBAP arrived at Priluki AB at noon on 17th April 1987, although some sources quote the date as 23rd or 25th April. One of the bombers was piloted by DA Deputy Commander Lev V. Kozlov, by then promoted to Lieutenant-General. The ferry flight went uneventfully; at Priluki, apart from the traditional welcoming committee and brass band, the airmen were 'welcomed' by a host of counter-intelligence corps officers whose job was to keep all information about the Tu-160 from leaking out.

On 12th May Kozlov made the first flight from Priluki; on 1st July this was followed by

184th GvTBAP Tu-160 '21 Red' is handled by a BelAZ-7420 airport tug after landing at Priluki AB. The aircraft is in the ultimate configuration with six intake blow-in doors per engine, a short fin bullet fairing and fin trailing edge ESM fairings. A sister ship is visible in the background.

the first sortie of a 184th GvTBAP crew captained by the regiment's CO Lt. Col. Vladimir Grebennikov. That same day Maj. Aleksandr S. Medvedev, the first commander of the first operational Tu-160 squadron, made his first solo flight. (Incidentally, in the 184th GvTBAP he was jokingly called *'ahs Medvedev'* (Ace Medvedev) – a pun on his initials, A. S.) By the end of the month Maj. Nikolay Stooditskiy, Maj. Valeriy Schcherbak and Maj. Vladimir Lezhayev had got their taste of flying the Tu-160. Interestingly, the same aircraft was used initially by all crews; it did not even taxi in to the flight line, and preparations for the next training flight (refuelling etc.) took place directly on the runway.

In late July 1987 (some sources state early August) a Tu-160 crew captained by Lt. Col. Vladimir Grebennikov performed the unit's first Kh-55SM launch with excellent results. The crew also included WSO Maj. Igor' Anikin and inspector pilot Lt. Gen. Lev V. Kozlov.

To speed up conversion training and conserve the bombers' service life a flight simulator was set up at Priluki AB. In order to make maximum use of the two aircraft delivered initially and train as many pilots as possible, it was standard operational procedure (SOP) at first to have the crews literally standing in line beside the runway, waiting for their turn to fly!

The airmen liked the Tu-160. The sparkling white bomber was very much a 'pilot's aeroplane', possessing excellent acceleration and

Above: A Kh-55MS cruise missile in long-range configuration is prepared for loading into the rear weapons bay of a Tu-160 at Priluki. The dark ring aft of the radome is a protective canvas wrap.

rate of climb (pilots said of the Tu-160 that it 'climbed by itself'). The aircraft also possessed good low-speed handling, which facilitated the landing procedure; at 260 km/h (161.5 mph; 140.5 kts) the minimum forward flight speed was even lower than the Tu-22M3's. The engines' total thrust was tremendous. On one occasion a Tu-160 even managed to become airborne and climb with the spoilers deployed by mistake; of course the climb was very slow at first but when the crew realised what was wrong and retracted the spoilers the aircraft *'shot up with such force that the crew almost punched through the seats with their backsides'*! The *Blackjack* featured an audio warning system and a stick-pusher which prevented grave piloting errors.

Dubbed 'The Pride of the Nation', the Tu-160 commanded respect from everyone who had to deal with it. Hence every due care was taken to ensure trouble-free operation. For instance, in the first months of operational service the crews were expressly forbidden from taxying to the holding point under the air-

Resplendent in its all-white finish, '21 Red' prepares to taxi at Priluki AB on a sunny day.

Above: A 1096th GvTBAP Tu-160 caught by the camera seconds before touching down at Engels-2 AB.

craft's own power in order to prevent foreign object damage (FOD) to the engines. The *modus operandi* was this: the engines were started on a special parking stand which had been painstakingly swept clean of all loose objects; then the bomber was towed to the runway with the engines running at ground idle, preceded by a string of soldiers picking up loose stones and twigs. As for the runway, the ground personnel did everything to keep it clean short of washing it with soap!

Speaking of nicknames, the Tu-160 is sometimes unofficially called **Belyy lebed'** (White Swan) because of its overall white finish and sleek lines. Also, the author of an article in *Nezavisimoye voyennoye obozreniye* (Independent Military Review, a weekly supplement to the *Nezavisimaya gazeta* daily, proposed officially naming the bomber *Il'ya Muromets II*. Supporting his idea, he alluded both to the pre-WWI four-engined Russian bomber designed by Igor' I. Sikorsky and to the US practice of re-using popular names (McDonnell F2H Phantom/F-4 Phantom II, Douglas C-74 Globemaster/C-124 Globemaster II/McDonnell Douglas C-17 Globemaster III etc.).

Combining the final stage of the trials with the service tests of the Tu-160 saved the Soviet state a lot of money. Still, the Tupolev OKB and the Air Force had to wrestle with the bomber's design flaws and operational problems for many years yet. Malfunctions occurred in virtually every single flight – primarily in the avionics which were a real pain in the neck. Luckily there were no serious accidents, mainly thanks to the multiple redundancy of the vital systems. The teething troubles and the generally hasty approach to the development of the Tu-160 caused the service evaluation period to drag on for several years. However, it should be remembered that the nation's leaders kept urging the Tupolev OKB and the Air Force to field the new bomber as soon as possible, which is why the Tu-160 entered service with lots of bugs still to be ironed out.

Problems with the powerplant were encountered regularly; engine starting was especially troublesome and the full authority digital engine control system could not cope with it. FADEC failures occurred in flight as well. On one occasion a Tu-160 captained by Maj. Vasin lost two engines at once. The petals of the NK-32's variable nozzles often failed. On the other hand, the *Blackjack* had adequate power reserves to maintain level flight and even take off with one engine inoperative (which came in most handy on an important occasion, as described later in this chapter).

One particular problem area was the air intakes whose aerodynamic imperfections created an annoying rasping noise and vibrations. The latter were downright dangerous, as they caused rivets to pop and fatigue cracks to appear. This defect was eradicated in due course by redesigning the forward sections of the inlet ducts (introducing a sixth blow-in door for each engine) and reinforcing the air intake leading edges.

The original design of the main landing gear units proved excessively complicated, leading to malfunctions (the bogies would get stuck halfway through rotation and so on). Reliability problems even caused 184th GvTBAP crews to refrain from retracting the gear for several months in 1988. This led the OKB to develop a simpler version which was introduced on the production line in Kazan' and retrofitted to some of the existing aircraft. Changes were also made to the hydraulic system.

Structural strength problems were encountered with the honeycomb-core panels utilised in the bomber's tail unit: the bonded panels would crack open at high speeds, causing severe vibration. One of the Tupolev OKB's own Tu-160s lost a good-sized fragment of the horizontal tail during a test flight; an identical incident occurred with a 184th GvTBAP aircraft captained by 'Ace Medvedev'. On another occasion a glassfibre section of the fin fillet broke away during a demonstration flight in Ryazan'. Eventually the Tupolev OKB was forced to ground the Tu-160 fleet and undertake an urgent redesign, reducing the horizontal tail span to 13.26 m (43 ft 6 in).

Now all existing Tu-160s had to have their stabilisers replaced with new ones, and an immediate problem arose: the all-movable stabilisers were too large to be carried internally by any transport aircraft in Soviet Air

A Tu-160 in high-speed cruise with the wings at maximum sweep. The erected wing fences are well visible.

Above: Tu-160 '03 Red' commences a late-afternoon take-off at Engels-2 AB, leaving the characteristic 'fox tail' wake of nitrous oxide behind. The special organic-base paint on the fuselage nose reduced unwanted electromagnetic emissions but gave the aircraft a rather untidy appearance.

Another view from Engels as Tu-160 '03 Red' taxies for take-off.

Above: For extreme high-altitude missions Tu-160 crews wear Baklan pressure suits with full-face pressure helmets similar to an astronaut's attire.

On most missions, however, the airmen wear ordinary flying suits and ZSh-7B 'bone dome' helmets.

Force service. Ferrying the bombers to Kazan' for a refit was ruled out for safety reasons. The solution was to carry the stabilisers externally. Thus a military IL-76 *sans suffixe* owned by the Kazan' aircraft factory (CCCP-76496, c/n 073410301, f/n 0806) was modified to carry Tu-160 horizontal tail assemblies from Kazan' to Priluki AB where they were replaced *in situ*. The unusual cargo was mounted atop the fuselage immediately aft of the wings on special struts; hence the aircraft was popularly known as the *tri**plahn*** (triplane). The first flight in this configuration took place on 30th October 1986; when the refit programme had been completed the *Candid-B* was reconverted to standard configuration.

One more problem (which, incidentally, affected the B-1 as well) quickly surfaced in service: with the wings in maximum sweep position (65°) the CG shifted so far aft that the bomber could easily tip over on its tail, and getting it back into normal position wasn't easy at all. Hence the wings had to be left at minimum sweep position (20°) on the ground, even though the Tu-160 required a lot more apron space in this configuration.

Some of the incidents were caused by pilot error. For instance, there were cases when the aircraft overran the runway because the brake parachutes were deployed too late. Slowing the *Blackjack* down was not easy due to the aircraft's high weight and hence inertia, and the pilots were unwilling to deploy the 'chutes too early for fear of losing face. Yet most problems were of a technical nature; for instance, the wheel brakes sometimes locked uncommandedly on take-off.

Much aggravation was caused by the state-of-the-art avionics, notably the Baikal ESM suite, 80% of which was housed in the aft fuselage. This area was subjected to serious vibration when the engines were running (particularly in full afterburner) and the vibrations shook the delicate electronics to bits. Tu-160 pilots caustically commented that they were lugging two tons of ballast for no purpose. By the spring of 1990 the ESM suite had worked up an acceptable reliability level, but still malfunctions did occur from time to time.

Another thing which caused a lot of problems for the ground personnel was the technical documents for the aircraft – or, to be precise, the lack of same.

Despite being easy to fly and stable in all flight modes, the Tu-160 nevertheless gave the aircrews many causes for criticism. For instance, the pilots were unhappy with the Zvezda K-36DL zero-zero ejection seat. Though a magnificent piece of engineering in itself (a fact proved by the many lives it saved), the K-36 was basically a fighter seat which turned out to be totally unsuited for the long missions of a strategic bomber. The same applied to the standard-issue flying suits designed for fighter pilots. The special protective helmets were in short supply for a long time, forcing several crews to take turns using the same set of helmets; as often as not the helmet was not the right size, creating an additional inconvenience for the crewman. Special heat-insulated rescue suits for over-water flights which made it possible to survive an ejection into ice-cold water were not available either (see below).

That was not the worst of it, however. The K-36DL seats could be adjusted lengthwise, and it turned out that ejection was impossible with the seat in certain positions. This was an extremely dangerous defect, and strictly speaking, an aircraft with such a defect should not be allowed to fly at all. For a long time the engineers at NPP Zvezda (NPP = na**ooch**no-proiz**vod**stvennoye predpri**yah**tiye – Scientific & Production Enterprise) tried to persuade the airmen not to slide the seats into the controversial position but eventually admitted that the situation was unacceptable and set about modifying the seat.

Flight deck ergonomics also gave cause for complaint. For instance, originally the main and back-up flight instruments were of different types, which was inconvenient; later the instrument panels were modified as recommended by service pilots to feature standardised main/back-up instruments.

To give credit where credit is due, the Tu-160's flight deck had a number of features enhancing crew comfort which had been unheard-of on Soviet bombers until then. These were a small nickel-plated oven for heating food and a toilet bowl. By comparison, on other bombers the crews had to chew cold meals and make do with a tin bucket, should they feel the urge to relieve themselves. There have been claims in the Russian press that the Air Force refused to accept the new Tupolev bomber for several months because it was not satisfied with the design of the toilet!

It should also be noted that at an early stage the Tupolev OKB had taken steps to facilitate the Tu-160's maintenance procedures as much as possible. For easy access the hydraulic systems components were arranged on the walls of the weapons bays, while the electric power distribution panels and fuse boxes were located in the wheel wells. Access to the engines was also reasonably good and the avionics racks in the flight deck and avionics bay were well designed. Still, the *Blackjack* turned out to be extremely 'labour-intensive', so to say, setting an unofficial Soviet Air Force record (of the unwanted kind) – it required 64 man-hours of maintenance per flight hour.

Preparing the Tu-160 for a sortie required between 15 and 20 ground support vehicles,

Above: Tu-160 '03 Red' is prepared for a sortie at Engels.

Six 184th GvTBAP Tu-160s lined up on the central taxiway of a dispersal area at Priluki.

73

A magnificent view of a *Blackjack* cruising over snow-covered countryside in a remote region of the Soviet Union with the wings at maximum sweep. The black circle aft of the flight deck is the window of the astro-inertial navigation system.

Above: A Kh-55SM cruise missile has just assumed cruise configuration after being released by Tu-160 '01 Red' during trials. The orange colour of the missile identifies it as an inert test round; the grey conformal tanks stand out clearly against the orange background.

An uncoded Tu-160 formates with the camera ship with the wings at minimum sweep. The elongated fairings just inboard of the engine nacelles are for the outer wheels of the landing gear bogies.

Above: Maintenance day at Engels. This view shows how the sliding flight deck windows can be used for evacuation on the ground.

Another view of a *Blackjack* cruising above the clouds over a mountainous landscape.

Above: A Tu-160 flanked by two Su-27 Flanker-Bs makes a flypast during one of the air fests at Moscow-Tushino.

Seen from an IL-78 tanker, a Tu-160 moves into position for refuelling.

Above: Tu-160 '05 Red' about to touch down at Engels past a line of Myasishchev 3MS-2 tankers.

1096th GvTBAP *Blackjacks* parked at Engels in 1994. The nearest aircraft ('04 Red') has a tarpaulin draped over the flight deck to save the glazing from being damaged by the ultraviolet radiation of the sun (this is SOP in the Russian Air Force). Note the open cowlings and the support placed under the tail.

Above: Tu-160 '05 Red' taxies in at Engels, with 1230th APSZ (Aerial Refuelling Regiment) Myasishchev 3MS-2 and Il'yushin IL-78M tankers (both quasi-civil and overtly military ones) in the background.

Another view of the same aircraft, showing the natural metal upper portions of the engine nacelles and the steel heat shields over the fuel transfer conduits to the rear fuselage fuel tank.

Above: A Tu-160 about to be demonstrated to a foreign delegation. Note the two Kh-55SM missiles and data tables beside the port wing. The flight crew stands beside, wearing their 'bone dome' helmets, which indicates that the day's programme included a demonstration flight.

some of which were perhaps unique for the type. These included three huge TZ-60 fuel bowsers (TZ = *toplivozaprahvschchik* – refuelling truck) consisting of a MAZ-537 or MAZ-543 eight-wheel tractor (popularly known as *Ooragahn*, Hurricane) with semitrailer, a special fuel nitrogenation unit, heavy-duty air conditioners mounted on KamAZ lorries for cooling the avionics, a minibus for the crew equipped with an air conditioner for the pressure suits etc. (Speaking of which, the NK-32 had a high nitrous monoxide content in its exhaust due to running on nitride fuel – after all, it was a military engine and had manners to match. Hence at full power the Tu-160 left a characteristic orange-coloured trail dubbed *lisiy khvost* (fox tail).)

The 'black men' (as Soviet Air Force technicians were nicknamed because of their black overalls) were confronted with other problems which were just as serious. When the engines were started the noise and vibration were horrendous, reaching 130 dB, which is 45 dB over safe limit determined by the medics; the TA-12 APU was not very quiet either, to say the least. To make matters worse there was an acute shortage of ear protectors, special vibration-damping boots and vibration protection belts. Instead of the usual AMG-10 or AMG-16 oil-based hydraulic fluid the Tu-160's hydraulic systems used a highly corrosive 7-50S-3 grade fluid. The lack of proper protective gear meant that the ground personnel could not do their job properly. Finally, some of the bomber's structural components and equipment items were difficult to repair and maintain.

Here it is worth quoting an article published in **Krahs**naya **zvez**dah (Red Star), the Soviet MoD daily, on 23rd July 1989. After describing the 'acute problems' experienced with the Tu-160, something they could do without, the pilots who had tested the bomber confided to the newspaper reporter, 'The aircraft won our hearts with its impressive capabilities. Still, you can tell that nobody was, so to say, snapping at the Tupolev OKB's heels. They hold a monopoly in this field!' Guards Col. Ye. Ignatov, the 184th GvTBAP's Chief of Maintenance, had a similar opinion: 'Eliminating the Myasishchev OKB was a mistake. We need competition!'

All problems notwithstanding, the 184th GvTBAP's 1st Squadron was fully ready for operational duty with the Tu-160 after an eight-month period. The unit did its best to overcome the learning curve; average Tu-160 utilisation rose steadily, reaching 100 flight hours per annum. Six-, ten- and twelve-hour missions were flown on a regular basis. Many pilots who transitioned to the *Blackjack* from the Tu-16 or Tu-22M3 said this was the best aircraft they had ever flown.

A weapons system of this complexity demanded a completely new approach to ground crew training. Suffice it to say that in the first days of Tu-160 operations in the 184th GvTBAP it took up to 36 hours (!) to prepare the aircraft for a mission. Gradually mission preparation times were reduced to an acceptable level thanks to the persistent work of the unit's tech staff and the immense help of the OKB's operations department.

The inevitable teething troubles accompanying the bomber's service introduction period were the subject of close attention from the aircraft industry. Up to 300 MAP representatives were on temporary detachment

A Tu-160 parked at Engels with canvas air intake covers and a tarpaulin cover over the flight deck to protect the glazing.

Above: A picture taken off a TV screen showing 184th GvTBAP Tu-160 '12 Red' flanked by a MiG-29 and an IL-78 at Kubinka AB on 12th August 1988 when it was demonstrated to US Secretary of Defense Frank C. Carlucci. This was the first time reasonably good pictures of the bomber were shown by the Soviet media.

to Priluki AB at any one time to resolve any problems on site; the defects discovered were quickly eliminated on both operational aircraft and the Kazan' production line. For instance, the NK-32's service life, which initially was a mere 250 hours, was tripled in due course. The number of auxiliary blow-in doors in the engines' inlet ducts was increased from five to six in order to prevent compressor stalls and their control system was simplified. Some honeycomb-core metal panels were replaced by composite panels of a similar design, saving weight and improving fatigue life.

At high speed the pointed fairing at the fin/stabiliser junction created vortices which caused dangerous vibrations ruining the ESM equipment in the aft fuselage. The problem was cured by shortening the fairing 50% and giving it a more bulbous shape. Late-production aircraft introduced rear vision periscopes for the navigator and WSO. These upgrades were effected on in-service Tu-160s by KAPO specialists at Priluki AB.

Some time after service entry the *Blackjack* received an avionics upgrade. Improvements were made to the long-range radio navigation (LORAN) system working with ground beacons. The navigation system was augmented by a self-contained celestial corrector which accurately determined the aircraft's position with the help of the sun and the stars; this was a real asset when flying over the ocean and in extreme northern latitudes. The navigators were pleased with the new PA-3 moving map display showing the aircraft's current position. A satellite navigation system with an accuracy margin of 10-20 m (33-66 ft) was also due for introduction on the Tu-160; it worked with several MoD satellites launched into geostationary orbit under a special government programme. The engineers also succeeded in debugging the software of the targeting/navigation avionics suite.

The Tupolev OKB devised and implemented a multi-stage programme aimed at making the *Blackjack* more stealthy. The air intakes and inlet ducts were coated with black graphite-based radar-absorbing material (RAM). The forward fuselage received a coat of special organic-based paint, some engine components were provided with special screens minimising radar returns, and a wire mesh filter was incorporated into the flight deck glazing to reduce electromagnetic pulse emissions which could give the bomber away. This filter also helped to protect the crew from the flash of a nuclear explosion.

The persistent and concerted efforts of the Air Force and the aircraft industry soon bore fruit; little by little the Tu-160 turned into a fully capable weapons system. The reliability of the aircraft in first-line service was improved no end. Still, many of the planned improvements never materialised, including the intended upgrade of the navigation/targeting suite. Unlike their American colleagues flying the B-1B, Tu-160 pilots never mastered ultra-low-level terrain-following flight, and far from all *Blackjack* crews were trained in IFR procedures which gave the aircraft its intended intercontinental range. The Kh-15S short-range ASM was never integrated, leaving the Tu-160 with the Kh-55 cruise missile as its only weapon.

Gradually, as the Soviet Air Force built up operational experience with the type, the

Even when the wings were at minimum sweep, a support was often placed under the aft fuselage for good measure to prevent the aircraft from falling over on its tail, as illustrated by '04 Red' at Engels. This was probably because the aircraft has an aft CG when the fuel tanks are empty.

Top and above: Front and rear views of a Tu-160 as it makes a flypast with the high-lift devices fully extended.

range of the Tu-160's missions expanded. The bombers went as far up north as the North Pole, sometimes even venturing across the Pole. The longest sortie – flown by a crew under Col. Valeriy Gorgol' who succeeded Vladimir Grebennikov as 184th GvTBAP CO in 1989 – lasted 12 hours 50 minutes; the bomber got within 450 km (279.5 miles) of the Canadian coast. NATO fighters often scrambled to intercept the Tu-160s over international waters; the first such occasion was apparently in May 1991 when a pair of Royal Norwegian Air Force/331 Sqn Lockheed F-16A Fighting Falcons from Bodø AB intercepted Tu-160 '17 Red' over the Norwegian Sea.

This is how Distinguished Military Pilot Aleksandr S. Medvedev ('Ace Medvedev'), who gained Russian citizenship after the break-up of the Soviet Union and subsequently became a Senior Inspector Pilot with the Combat Training Department of the 37th VA of the Russian Armed Forces Supreme Command (the former Long-Range Aviation), describes the *Blackjack*'s service introduction period:

'We first got acquainted with the Tu-160 about a year before the first bombers arrived at Priluki. We visited the factories where the aircraft proper, its engines and avionics were manufactured to study the hardware first hand. After we had received the Tu-160s flight and combat training got under way. The Cold War was far from over then, and we were tasked with mastering the new aircraft as quickly as possible. Note that we had to do this in the course of service tests, which we had to perform instead of the factory pilots. This is why we got up to 100 systems malfunctions per flight in the early sorties. As we grew familiar with the machine the number of failures decreased, and in due time we came to trust the aircraft.

Where did we fly, you ask? We departed Priluki in the direction of Lake Onega (near Petrozavodsk in Karelia, north-western Russia – Auth.), thence we flew to Novaya Zemlya Island ('New Land' – Auth.), the Franz-Josef Land archipelago and onwards, heading towards the North Pole. The guys back at home used to joke that 'the Cowboy (the nickname of one of the Tu-160s – Auth.) is studying the States'; indeed, the 'potential adversary' was within easy reach for us. However, over the Pole we would turn and head towards Tiksi (on the Laptev Sea coast – Auth.); all the more reason to do so because the terrain there was similar to the coasts of northern Canada. Next thing we would overfly Chelyabinsk (in the southern part of the Urals mountain range – Auth.), Lake Balkhash and the Caspian Sea, heading across the North Caucasus back towards Priluki. There was also a route from Priluki to Lake Baikal and back again; I used to fly it with Vladimir Grebennikov. Missiles were launched against targets on a range in Kazakhstan; we would enter the designated launch zone and say goodbye to our missiles. Later we also practiced missile attacks at target ranges in other parts of the USSR.'

On some occasions the Tu-160s were escorted by Sukhoi Su-27P *Flanker-B* interceptors of the Air Defence Force's 10th Army operating from bases near Murmansk and on Novaya Zemlya Island. Overwater missions were flown in pairs because the presence of a wingman (or flight leader, depending on what aircraft in the pair you flew) added psychological comfort; if one of the crews ran into trouble and had to ditch or eject over water, the other crew could radio for help, indicating the co-ordinates of the crash site. This was important because Tu-160 crews had nothing more than DSP-74 life jackets for such an

Updated Tu-160s of the 184th GvTBAP in the revetments at Priluki AB. The ZiL-131 under the port wing is probably an oxygen or nitrogen charging vehicle, while the other one is a UPG-300 power cart.

Above: Tu-160 '18 Red' was among the advanced Soviet Air Force aircraft displayed to the political and military leaders of the newly-formed Commonwealth of Independent States at Machoolischchi AB near Minsk, Belorussia, on 13th February 1992. Here the show has yet to begin and the flight deck is still under tarps.

The day before the morning after. '18 Red' is is washed and polished at Machoolischchi on 12th February 1992 prior to the following day's demonstration.

Above: 1096th GvTBAP Tu-160 '03 Red' is prepared for a practice mission at Engels-2 AB. The vehicle in the foreground is a TZ-60 fuel bowser (pulled by the older MAZ-537 in this case).

Another angle on the same scene, giving a good view of the wing joint. Note that the spoilers are deployed.

Above: A typical publicity photo of a *Blackjack* flying over thick overcast.

Early morning scene at Engels-2 AB. This view shows well the rear ESM antennas and APP-50 flare dispensers. Note the Tu-22KD and Tu-22KPD missile carriers in the background.

Above: A poor but interesting shot of a pair of *Blackjacks* in echelon port formation. The aircraft have a well-worn appearance.

Lit by the bright sun and pictured against the violet skies of the stratosphere, the Tu-160 is an impressive sight.

Above: This picture of '03 Red' as it banks away from the camera was taken from the open entry door of a Tu-134UBL!

Another air-to-air of the *Blackjack* taken in the same fashion.

emergency; the VMSK-4 heat-insulated waterproof rescue suit (***vod**opronit**say**e-myy mor**skoy** spa**sah**tel'nyy kos**tyoom***) was a privilege of Naval Aviation pilots. This situation was caused by the lumbering bureaucratic machine responsible for materiel supplies in the Armed Forces. Fortunately no such emergencies have occurred to date... knock on wood.

One of the reasons why the new strategic bomber became operational very quickly was the targeting and navigation system's high degree of automation which reduced the WSO's workload. The WSO was the key member of the crew during missile launch. As already noted, the Kh-55SM cruise missile was guided to the target by a pre-entered programme featuring a digital route map; therefore the WSO's duties basically boiled down to accurately guiding the aircraft to the launch point, monitoring the missiles' systems status and pushing the release button at the right time. The missile was then ejected downwards by a pneumatic catapult built into the rotary launcher; at a safe distance from the aircraft its wings and tail surfaces unfolded, the engine was deployed and fired up, and the missile followed its intended course. Meanwhile the launcher rotated, bringing the next missile into position for release.

All practice launches of the Kh-55SM took place at the GK NII VVS target range in Kazakhstan and were supported by 'aircraft 976' airborne measuring and control stations. Live weapons training with the new cruise missile proceeded on a much wider scale than even with the Tu-22M3 armed with the proven Kh-22N (AS-4 *Kitchen*) ASMs. One of the Tu-160s – the one known in the regiment as the 'Cowboy' (presumably because it had two 'six-shooters' inside! – Auth.) launched no fewer than 14 missiles. Maj. I. N. Anisin, the 184th GvTBAP's intelligence section chief, was one of the top-scoring WSOs – which was understandable enough, as he had to know all about the unit's potential targets.

By the end of 1987 the 184th GvTBAP had increased its Tu-160 fleet to ten aircraft; nevertheless, the regiment stuck to its Tu-22M3s and Tu-16P *Badger-J* ECM aircraft to maintain combat readiness during the transition period. Later, as the number of *Blackjacks* grew, the older types were progressively transferred to other units; some of the Tu-16s were scrapped on site to comply with the Conventional Forces in Europe (CFE) treaty limiting the number of combat aircraft a unit was authorised to have.

The strategic bombers themselves were governed by a different treaty, the Strategic Arms Reduction Talks (START) treaty. A group of US inspectors was to arrive at Priluki AB in order to monitor the number of strategic aircraft, and a bungalow was specially built for them near the runway and hardstands.

As the economic downturn caused by Mikhail S. Gorbachov's *perestroika* got worse, Tu-160 production and delivery rates slowed down somewhat; by late 1991 the 184th GvTBAP had 21 *Blackjacks* in two squadrons. At the beginning of that year the unit's 3rd Squadron had received a small number of Tu-134UBLs; until then the type had been operated solely by the military flying schools in Orsk and Tambov. The *Crusty-Bs* were used for lead-in and proficiency training, saving the bombers' service life and helping to avoid unnecessary breakdowns and costly repairs.

The Western intelligence community maintained a close interest in the Tu-160, especially its mission equipment, long before the aircraft entered service. The Soviet counter-intelligence service (the notorious KGB) was extremely alarmed to discover a self-contained signals intelligence (SIGINT) module near Priluki AB in the spring of 1988. Disguised as a tree stump, the thing monitored and recorded air-to-ground radio exchanges and other signals emitted by the aircraft operating from the base. It was never ascertained who had planted the module, but countermeasures were not slow in coming: operational Tu-160s were provided with 'nightcaps' made of metal-coated cloth which were placed over the nose on the ground to contain electromagnetic pulses. As a bonus, these covers protected the ground personnel from harmful high-frequency radiation when the avionics were tested on the ground.

On the other hand, *perestroika*, *glasnost'* (openness) and the new domestic and foreign policies initiated by Gorbachov removed the pall of secrecy from the Soviet Armed Forces and defence industry, making a lot of previously classified information available via the mass media. Also, in addition to military delegations from allied nations the Soviet Union began inviting high-ranking military officials from countries previously regarded as 'potential adversaries'. New Soviet military hardware was demonstrated to Western experts both at home and abroad during major international airshows and defence trade fairs.

The West got its first close look at the *Blackjack* on 12th August 1988 when Frank C. Carlucci, the then US Secretary of Defense, visited Kubinka AB near Moscow during his Soviet trip. Kubinka had a long history as a display centre where the latest military aircraft were demonstrated to Soviet and foreign military top brass. The aircraft shown to Mr. Carlucci included a Mikoyan MiG-29 *Fulcrum-A* tactical fighter, a Mil' Mi-26 *Halo* heavy-lift helicopter, an IL-78 *Midas-A* tanker/transport – and a 184th GvTBAP Tu-160 coded '12 Red'.

In an unprecedented show of openness, the Secretary of Defense was allowed to inspect the weapons bays, the flight deck and other details of the *Blackjack*. The US delegation was accompanied by TV crews and press photographers, and soon the first reasonably good pictures of the Tu-160 were circulated in the world media. Also, some performance figures were disclosed for the first time, including an unrefuelled range of 14,000 km (8,695 miles).

As is usually the case on such occasions, a flying display was staged; it included two more Tu-160s which were parked elsewhere on the base. When the bombers (captained by Vladimir Grebennikov and Aleksandr Medvedev) were due to taxi out for their demonstration flight, a single engine would not start on each aircraft (!). Realising the embarrassment that would result if the demo flight was cancelled for this reason, the VVS top command authorised the crews to take off on three engines – which is exactly what they did. The flights went well, thanks as much to excellent airmanship as to the *Blackjack*'s good flying qualities.

Nevertheless, the fact that only three of each bomber's engines were emitting the characteristic orange efflux did not escape the attention of the US Air Force representatives, who demanded an explanation. Worried though he was about the situation, Long-Range Aviation Commander Col. Gen. Pyotr S. Deynekin answered with a straight face that the Tu-160's engines had several operational modes, not all of which were characterised by a smoke trail.

It is not known whether the Americans believed him; indeed, it would appear hardly improbable that ordinary service pilots would run the risk of taking off with one engine dead. However, even if they had guessed the truth and had the grace to say nothing, they surely acknowledged that the Soviet pilots were real pros.

Another embarrassing situation showing that the Tu-160 still had a few bugs arose when Frank Carlucci climbed into the flight deck. As he moved about in the confined space he hit his head on a carelessly positioned circuit breaker panel (which the witty Soviet airmen promptly dubbed 'the Carlucci panel'). Incidentally, almost every person making his first visit inside the *Blackjack* bumps his head on it!

After this event the bomber became a frequent participant of both similar shows for foreign military delegations at Kubinka and the airshows at Moscow-Tushino and Zhukovskiy for the benefit of the general public. For instance, the French Defence Minister Jean-Pierre Chevènement examined Tu-160 '16 Red' at Kubinka AB in March 1989. Three months later, on 13th June, another 184th

Above: Ukrainian Air Force/184th GvTBAP Tu-160 '14 Red' on display at Poltava AB in September 1994 during the festivities marking the 50th anniversary of the shuttle raids against Germany. USAF bombers used this particular airfield for refuelling and re-arming for another attack on German targets on the way home.

Another view of '14 Red' at Poltava in company with a visiting Boeing B-52H Stratofortress, a Rockwell B-1B and a McDonnell Douglas KC-10A Extender tanker. This was one of the few Ukrainian *Blackjacks* to wear full UAF insignia with St. Volodimir's trident. It was scrapped in 1999 with less than 100 hours on the clock.

GvTBAP aircraft ('21 Red') was shown to Adm. William Crowe, Chairman of the US Joint Chiefs of Staff, at the same location. Climbing down the steep ladder from the flight deck (and presumably avoiding the famous panel), Crowe described the Tu-160 as 'a world class aircraft'.

The Tu-160's public debut took place on 20th August 1989 when one of the development aircraft from Zhukovskiy made a low pass over Tushino airfield as part of the annual Aviation Day display. The first known display in Zhukovskiy was on 18th August 1991 (one day before the failed hard-line Communist coup which brought an end to the Soviet Union's existence) – also on Aviation Day which is celebrated on the third Sunday in August. The huge territory of the Flight Research Institute was still off limits to the public then, but spectators sitting on vantage points near LII's perimeter fence could see the entire engine starting procedure and take-off, whereupon the aircraft came back to make a low-level pass over the temporary grandstand on the bank of the Moskva River.

On 13th February 1992 a 184th GvTBAP Tu-160 coded '16 Red' was demonstrated to top-ranking military officials and the leaders of the CIS states at Machoolischchi AB near Minsk along with other advanced military aircraft. Curiously, the data plates for the exhibits were carefully draped with black cloth to hide the "top secret" figures from prying journalists (who arrived in force) and unveiled only for the visiting VIPs. Obviously *glasnost'* still had a long way to go in 1992!

On 11th-16th August 1992 LII hosted the first fully-fledged Moscow airshow, MosAeroShow-92, where an unpainted Tu-160 development aircraft ('29 Grey') took part in the flying display. The static park included one more of the OKB's *Blackjacks*, '63 Grey' (c/n 84704217, f/n 0401), which the visitors were able to inspect at close range. Since then this particular aircraft has been an invariable participant of all subsequent Moscow airshows – MAKS-93 (31st August – 5th September 1993), MAKS-95 (22nd-27th August 1995), MAKS-97 (19th-24th August 1997), MAKS-99 (17th-22nd August 1999) and MAKS-2001 (14th-19th August 2001). The reason is probably that of all company-owned *Blackjacks* it was in the best condition. The numerous visitors and press were also able to see (and covertly photograph) several other Tu-160s on the Tupolev OKB hardstand some 200 m (660 ft) from the static display, but these were in various states of disrepair.

In September 1994 media representatives and aviation experts had the opportunity to get a close look at the Tu-160 at Poltava AB, in the Ukraine. The event marked the 50th anniversary of the shuttle raids against Nazi Germany when USAF Boeing B-17 Flying Fortresses and Consolidated B-24 Liberators landed at Poltava after unloading their bombs on German cities. One more open doors day was held at Priluki AB in February 1995.

The grand military parade at Poklonnaya Gora in Moscow on 9th May 1995 on occasion of the 50th anniversary of VE-Day featured a large Air Force component. The first fixed-wing aircraft in the flypast (after a quartet of Mil' Mi-8T *Hip* helicopters carrying flags) was Tu-160 '06 Red' (named *Il'ya Muromets*) accompanied by four MiG-29s of the Russian Air Force's *Strizhi* (Swifts) aerobatic team from Kubinka. The bomber, which was captained by none other than Pyotr S. Deynekin, had flown in non-stop from Engels-2 AB near Saratov via Kozel'sk, Yookhnov, Vereya and Kubinka where it was joined by the fighter escort. This is how Col. Gen. Deynekin himself describes the event:

184th GvTBAP Tu-160 '10 Red' at Priluki in post-Soviet times. Most Ukrainian Tu-160s had no national insignia, sporting greyish pentagons where the Soviet stars had been sloppily painted out.

'By producing the Tu-160 strategic aircraft in quantity Russia once again proved its status as a great aviation power and the birthplace of heavy aircraft construction. During the First World War, when the airmen of the West fearlessly buzzed about in single-engined Voisins, Blériots, Nieuports and Farmans, the Russians conquered the skies of war in four-engined aeroplanes bearing the proud name of an epic hero, Il'ya Muromets. This is exactly how we christened the flagship of the 9th May 1995 flypast in Moscow, reviving an old tradition. The legend was written on the Tu-160's side in large Cyrillic characters (Deynekin means 'in ancient Cyrillic script' – Auth.) – why not, since there was ample space! On the day of the jubilee not only the heads of state but also military attachés from 40 foreign countries were in attendance at Poklonnaya Gora. After the flypast many of them said that the Russian Air Force had professionally demonstrated to them the structure of the air arm's massed first echelon. And well they might! The errors in timing as we approached the guest grandstand was just seven seconds! Mind you that our Tu-160 crew had taken off from Engels airbase on the Volga River, while the other 79 [aircraft involved in the flypast] took off from 14 other airfields. The pilots flying the aircraft taking part in the parade included the C-in-C of the Air Force, the C-in-C of the Air Defence Force, the Commanders of the Army Aviation, Naval Aviation, Tactical Aviation, Long-Range Aviation and the military airlift arm. Eight Colonel-Generals in all! Nothing like this had been seen at flypasts even in the days of Vasiliy Stalin! (Vasiliy I. Stalin, the Soviet leader's son, commanded the Moscow Defence District air arm in the late 1940s and was a huge supporter of air parades, often participating in them personally – Auth.)

We completed the flying display and landed at our respective bases without anything untoward that time. We were even in time for the official reception in the Kremlin...

...You should have seen us fly during combat readiness checks! The entire unit would fly in the daytime and at night, in any weather! All the pilots – both captains and co-pilots – were strictly Pilots 1st Class in those days.' (Pilot 1st Class is another official grade.)

The next time the general public was treated to a demonstration of the Tu-160 was 28th July 1996, the day when the Russian Navy celebrated its 300th anniversary. The festivities in St. Petersburg, the birthplace of the Russian Navy, included a spectacular flypast over the Neva River featuring, among other things, a Tu-160 escorted by four MiG-29s of the Strizhi team. Piloted by Lt. Col. Aleksey Serebryakov, the Blackjack made a high-speed pass at about 50 m (160 ft) with the wings at maximum sweep.

It was not long before the Tu-160 demonstrated its high performance for the world to see. In October 1989 and May 1990 Soviet Air Force crews established an impressive series of world speed and altitude records. The Tupolev OKB duly submitted the data to the Fédération Aéronautique Internationale (FAI) but so far the author has found no confirmation they have been officially recognised. As of this writing the Tu-160 holds 44 Class C-1s/Group 3 records.

The collapse of the Soviet Union and the division of its assets between the new independent states had a detrimental effect on one of the world's most potent offensive weapons systems. On 24th August 1991 the Ukrainian parliament issued a decree placing all military

World records claimed by the Tupolev OKB for the Tu-160

Date	Crew	Description	Value	No. of records
31-10-1989	L. V. Kozlov M. I. Pozdnyakov V. N. Neretin S. N. Mart'yanov	Altitude with payload 0/1,000/2,000/5,000/10,000/15,000/20,000/25,000/30,000 kg (0/2,204/4,409/11,022/22,045/33,070/44,090/55,115/66,140 lb)	13,894 m (45,584 ft)	9
		Speed with 15,000/25,000-kg (33,070/55,115-lb) payload over a 1,000-km (621-mile) closed circuit	1,731.4 km/h (1,075.4 mph; 935.89 kts)	2
		Payload lifted to 2,000 m (6,560 ft)	30,471 kg (67,175 lb)	1
		Speed with payload over a 1,000-km (621-mile) closed circuit 0/1,000/2,000/5,000/10,000/15,000/20,000/25,000/30,000 kg (0/2,204/4,409/11,022/22,045/33,070/44,090/55,115/66,140 lb)	1,731.4 km/h (1,075.4 mph; 935.89 kts)	9
15-5-1990	L. V. Kozlov V. P. Rudenko S. N. Mart'yanov	Speed with payload over a 1,000-km (621-mile) closed circuit 0/1,000/2,000/5,000/10,000/15,000/20,000/25,000/30,000 kg (0/2,204/4,409/11,022/22,045/33,070/44,090/55,115/66,140 lb)	1,726.9 km/h (1,072.6 mph; 933.45 kts)	9
22-5-1990	N. Sh. Sattarov A. S. Medvedev P. P. Merzlyakov	Speed with payload over a 2,000-km (1,242-mile) closed circuit 0/1,000/2,000/5,000/10,000/15,000/20,000/25,000/30,000 kg (0/2,204/4,409/11,022/22,045/33,070/44,090/55,115/66,140 lb)	1,195.7 km/h (742.6 mph; 646.32 kts)	9
24-5-1990	V. I. Pavlov V. P. Selivanov F. A. Ivlev	Speed with payload over a 5,000-km (3,105-mile) closed circuit 0/1,000/2,000/5,000/10,000/15,000/20,000/25,000/30,000 kg (0/2,204/4,409/11,022/22,045/33,070/44,090/55,115/66,140 lb)	920.95 km/h (572.0 mph; 497.81 kts)	9
28-5-1990	S. D. Osipov N. N. Matveyev A. S. Tsarakhov	Speed with payload over a 5,000-km (3,105-mile) closed circuit 0/1,000/2,000/5,000/10,000/15,000/20,000/25,000/30,000 kg (0/2,204/4,409/11,022/22,045/33,070/44,090/55,115/66,140 lb)	1,017.8 km/h (632.17 mph; 550.16 kts)	9

Above: The nose of Tu-160 '33 Red' at Priluki, with the obligatory fire extinguisher 'parked' in front of the aircraft (also standard operational procedure at Soviet air bases). Note the nine star markers under the captain's window marking successful missile launches.

The tail of '10 Red' with freshly applied Ukrainian shield-and-trident insignia at Priluki AB.

'10 Red' was displayed at one of the open doors days at Priluki AB. A curious aspect of such events in the CIS is that the spectators are usually allowed to get close enough to touch the aircraft – a blessing for spotters but a nuisance for photographers as the exhibits are crawling with people (sometimes all too literally).

units and installations on the territory of the former Ukrainian Soviet Socialist Republic under Ukrainian control. The Ukrainian Ministry of Defence was formed that same day. The fledgling Ukrainian Air Force (*Vo-yenno-povitrayny seely Ookrayiny*) inherited the 19 Tu-160s of the 184th GvTBAP in Priluki – for which it had no use. These aircraft absolutely did not fit into the republic's military doctrine, since the Ukraine had proclaimed non-nuclear status.

In February 1992 the Russian Federation's first President Boris N. Yel'tsin signed a directive terminating production of the Tu-95MS *Bear-H* missile carrier in Samara. The document also envisaged the possible halting of Tu-160 production, provided that the USA stop production of the Northrop B-2 Spirit bomber; however, this initiative was not treated favourably in Washington. Besides, since the only operational Tu-160 unit was stationed outside Russia, the break-up of the Soviet Union effectively left Russia without new-generation strategic aircraft. Thus, costly though the bomber was, the *Blackjack* production line remained open for the time being.

New-build aircraft were now delivered to the 1096th *Sevastopol'skiy* GvTBAP at Engels-2 AB which was part of the 37th VA/ 22nd *Donbasskaya* GvTBAD (*tyazhelobom-bardeerovochnaya aviadiveeziya* – Guards heavy bomber division, = bomber group (heavy)). It should be noted that the original plans – as far back as the design stage – had been to deliver the first production Tu-160s to Engels, while the regiment in Priluki was then regarded as a reserve unit. Before re-equipping with the Tu-160 the 1096th GvTBAP had operated Myasishchev 3MD *Bison-C* four-jet subsonic bombers; these were supported by the co-located 1230th APSZ (*aviapolk samo-lyotov-zaprahvschchikov* – aerial refuelling regiment, = aerial refuelling wing) operating Myasishchev 3MS-2 *Bison-B* tankers.

The 1096th GvTBAP took delivery of its first brand-new Tu-160 on 14th February 1992 (some sources say 16th February); by May this number had increased to three aircraft. A large proportion of the 184th GvTBAP's flight and ground crews gave up their positions at Priluki and arranged a transfer to Engels in order to serve in the Russian Armed Forces, refusing to swear allegiance to the Ukraine. (Speaking of which, a total of 720 Air Force servicemen stationed in the former Ukrainian SSR joined the Russian Air Force in 1992-93.) A Tu-160 squadron was formed at Engels-2 AB. At the same time major reconstruction and upgrade work got under way at the base; all ground support equipment associated with Tu-160 operations, the flight simulator and a lot of other materiel had been left behind at Priluki AB and the 1096th GvTBAP had to start from scratch.

On 29th July 1992 Lt. Col. Aleksandr S. Medvedev, one of the ex-184th GvTBAP pilots who had moved to Russia, performed the Russian Air Force's first *Blackjack* flight. Thus Russia demonstrated *urbi et orbi* that it was still a strong aviation power that was to be reckoned with. On 22nd October 1992 Lt. Col. Anatoliy Zhikharev, the 1096th GvTBAP's CO, successfully launched a Kh-55SM cruise missile against a practice target; an identical mission was flown the following day by a crew under Lt. Col. A. V. Malyshev.

In early 1993 the unit received its fourth new-build Tu-160. This was clearly insufficient to maintain a fully combat-capable unit. To bolster Russia's operational *Blackjack* force it was suggested that the six Tu-160s owned by ANTK Tupolev and LII should be handed over to the 1096th GvTBAP, even though they were high-time airframes; however, the plan never materialised.

Thus, defying all difficulties caused by the chaos of the early post-Soviet years (fuel and spares shortages and so on), the Russian Air Force's 37th VA managed to keep at least a degree of combat readiness. Even in 1992, which was the hardest year, Long-Range Aviation pilots tried to maintain their class ratings by flying 80 to 90 hours per annum, which was twice the number of hours flown by Tactical Aviation (fighter and fighter-bomber) pilots!

In May 1993 Russian Air Force Tu-160s participated in Exercise *Voskhod-93* (Sunrise-93) which involved a rapid reaction scenario in the event of a sudden threat to national security. The bombers' long range enabled them to deploy rapidly to the Far East to cover one of the key defence areas, bolstering a group of Su-24M *Fencer-D* tactical bombers and Su-27 fighters deployed to the area. This time the *Blackjack* crews had to make do with simulated missile launches because there were no suitable target ranges in the Transbaikalian Defence District.

A live launch of an extended range version of the Kh-55SM fitted with conformal fuel tanks was performed in the course of strategic nuclear forces exercise which took place on 21st-22nd June 1994; the exercise was inspected by President Boris N. Yel'tsin himself. The missile destroyed a practice target at the Koora target range on the Kamchatka Peninsula, Far Eastern Defence District; other targets at the same range were successfully destroyed by RS-12 ***Topol'*** (Poplar; NATO code name SS-20) ICBMs and missiles launched by a North Fleet *Taifoon* (Typhoon) class nuclear submarine. (Of course, all the missiles were equipped with conventional warheads for practice purposes.)

Back in the Ukraine, the collapse of the Soviet Union had no dramatic effect on the daily activities of the 184th GvTBAP at first. However, in the spring of 1992 the Ukrainian government began administering the oath of allegiance to the military units stationed in the republic, and an exodus of personnel ensued. The 184th GvTBAP's turn was on 8th May 1992, but only about 25% of the flight crews and 60% of the ground personnel took the oath; the regiment's CO Col. Valeriy Gorgol' was the first to do so. The Ukrainian Air Force also appropriated the 409th APSZ at Uzin AB near Kiev which operated virtually the Soviet Air Force's entire fleet of IL-78 *sans suffixe* (*Midas-A*) tanker/transports. Most Ukrainian military aircraft received the new national insignia in the form of blue/yellow roundels on the wings and a blue shield with a yellow trident on the tail (the IL-76MD transports and IL-78s were a notable exception, remaining civil-registered). On the Tu-160s at Priluki, however, in most cases the old Soviet star insignia were simply painted out and the new insignia never applied.

Dire though the situation in the Russian Air Force was in the early 1990s, the Ukrainian military airmen were even worse off. The units operating heavy aircraft, the ones most difficult and expensive to maintain, were the hardest hit. Combat training had to be curtailed at once; the Ukraine had no target ranges of its own, and plans to establish a combat and conversion training centre for the Ukrainian heavy bomber crews never materialised. Furthermore, ANTK Tupolev and the Kazan' aircraft factory (which had delivered the bombers with a ten-year warranty) were no longer providing product support. Fuel and spares shortages, coupled with the exodus of qualified cadre, quickly grounded part of the bomber fleet. The severance of traditional ties within the former Soviet Union was a prime cause; for instance, the special IP-50 engine oil for the NK-32 turbofans was produced only in Azerbaijan, the landing gear wheels were manufactured in Yaroslavl' and the engines in Samara (formerly Kuibyshev). As systems components ran out of service life and no replacements were forthcoming, the 184th GvTBAP's maintenance department was forced to cannibalise some of the aircraft for spares to keep the others flyable. Still, by mid-1994 these unpopular measures could no longer save the day – the unit was left with only a handful of pilots qualified to fly the Tu-160 and they had the opportunity to fly only four or five times a year. The drop in flying hours per annum and the resulting lack of proficiency led to a higher number of malfunctions; the unit's CO Valeriy Gorgol' had to cope with the worst one in May 1993 when one of the main gear units jammed halfway through extension. Occasionally a single Tu-160 would take off from Priluki to participate in an airshow or a military parade. The bottom line is that the small Russian Air Force complement of five *Blackjacks* was more potent than the 21 Ukrainian examples of the type!

Russia and the Ukraine were at odds over a number of political and military issues, including the *Blackjacks*. Almost immediately after the break-up of the Soviet Union the Russian government started negotiating the purchase of the Tu-160s left behind at Priluki AB and the Tu-95MSs operated by the 1006th TBAP at Uzin AB. In March 1993 V. Zakharchenko, the advisor of the Ukrainian military attaché in Moscow, stated that 'the Ukrainian Armed Forces have no missions which these aircraft could fulfil'. V. Antonets, the then Commander-in-Chief of the Ukrainian Air Force, shared this opinion, saying in an interview at Priluki on 15th February 1995 that the crisis affecting the national economy made it impossible to adequately maintain the Ukrainian Tu-160s and therefore the Ukraine was interested in selling them to Russia.

Starting in 1993, the possible sale of the Tu-160s was brought into discussion more than 20 times but the parties could not agree about the price. The Ukraine's asking price was US$ 75 million per aircraft; Moscow tried to bargain for better terms again and again but Kiev stood firm on 75 million. Then Russia proposed exchanging the bombers for tactical aircraft and spares for same, but the Ukraine was not interested. Meanwhile the condition of the bombers, each of which required more than US$ 1 million per year in maintenance costs, slowly deteriorated. (Given the tensions between Moscow and Kiev, many people believed that the Ukraine was intentionally sabotaging the deal and would sooner let the bombers rot away than let the Russians have them!)

At the same time the US State Department started putting pressure on Kiev, demanding that the Ukraine comply with the START-2 treaty which required the Soviet Union to dismantle its strategic bombers not later than 4th December 2001.

Finally, in 1998 the Ukraine began scrapping its Tu-160s which the US military were so worried about. The USA assigned US$ 8 million for the destruction of these aircraft under the Co-operative Threat Reduction Program (also known as the Nunn-Lugar Program after Senators Samuel Nunn and Richard Lugar who got it through Congress). On 16th November 1998 the first of the Ukrainian Tu-160s to be scrapped, '24 Red', was ceremonially broken up at Priluki AB; the aircraft, which was manufactured in 1989, had logged 466 hours total time. This 'grand occasion' (for some people, that is) was attended by Senators Richard Lugar and Carl Levine. The work was supervised by the American aerospace company Raytheon.

The second *Blackjack* to be disposed of was '14 Red', one of the few to actually receive Ukrainian Air Force markings. This aircraft, which gained fame in the aviation community after being displayed at Poltava during the shuttle raid 50th Anniversary festivities in late September 1994, was built in 1991 and had less than 100 hours total time on it! It was totally broken up in November 1999.

In October 1999, however, the Russian media brought good news: the Ukraine had finally agreed to hand over eight Tu-160s and three Tu-95MSs to Russia to offset its outstanding debt for Russian natural gas deliveries. At the end of the month Vladimir V. Putin, the then Prime Minister of Russia, signed a directive approving a bilateral agreement under which the Ukraine would transfer eight *Blackjacks*, three *Bear-Hs* and more than 575 cruise missiles (including Kh-55SM air-launched cruise missiles), plus ground support equipment to Russia.

On 20th October 1999 a group of Russian Air Force/37th VA specialists led by Long-Range Aviation Deputy Commander Maj. Gen. Pyotr Kazazayev arrived in the Ukraine to take charge of the newly-acquired bombers. It would seem Kiev was regretting the deal it had made; for three days the Ukrainian Customs service would not let the bombers leave the country, demanding some

Above: Ex-Ukrainian (and unmarked) Tu-160 '16 Red' takes off from Priluki on the way to its new home at Engels.

'25 Red' and '26 Red' with no national insignia on the taxiway of a dispersal at Priluki AB; the tails of two further *Blackjacks* in full Ukrainian insignia (including '15 Red') are visible inside the revetments. Most of these aircraft were destined for the breakers.

Escorted by four MiG-29s of the *Strizhi* **display team ('40 White', '42 White', '43 White' and '46 White'), Russian Air Force/121st GvTBAP Tu-160 '06 Red'** *Il'ya Muromets* **passes over Poklonnaya Gora in Moscow on 9th May 1995 during the VE-Day 50th Anniversary parade.**

'01 Red' makes a flypast during one of the open doors days at Kubinka AB, this time escorted by a quartet of Su-27s of the resident *Roosskiye Vityazi* (Russian Knights) display team.

Above: Another airshow performance in company with *Flanker-Bs*, this time service aircraft in standard camouflage. The No.1 right-hand wingman is an early-production Su-27, as indicated by the dark green radome (as distinct from the white radomes on the other fighters).

An uncoded *Blackjack* makes a flypast at Moscow-Tushino in company with two Su-27s during the 1998 Aviation Day air fest.

Above: In 1995 the Russian Air Force started naming its Tu-160 fleet (then numbering only six aircraft), a practice that soon spread to other types. This is *the other Il'ya Muromets*, '05 Red' – the one which was later renamed *Aleksandr Golovanov*.

'01 Red' is named after the famous test pilot Mikhail M. Gromov whose name the Flight Research Institute in Zhukovskiy also bears. Unlike '05 Red' and '06 Red', this aircraft combines a Russian flag fin flash with the traditional red star.

'07 Red' *Aleksandr Molodchiy* (f/n 0802) is the newest Tu-160 in the Russian Air Force. It is named after a famous bomber pilot of the Second World War.

Tu-160 '11 Red' is named after Aleksandr Novikov, a Soviet Air Chief Marshal of the 1940s.

Above: '04 Red', named after wrestler Ivan Yarygin, lacks the colourful nose and tail trim. It is seen here landing at Ryazan'. Note the grey-painted Tu-134UBL pilot trainers with the characteristic pointed nose and red lightning bolt and the red-tailed Tu-134Sh-1 navigator trainers of the 43rd Combat & Conversion Training Centre.

'10 Red', one of the ex-Ukrainian *Blackjacks* transferred to the 121st GvTBAP and thus saved from the breaker's torch, is the opposite case: it has the nose and tail flashes but no name as yet. Note the nose-down position of the main gear bogies in no-load condition.

Above: '11 Red' *Vasiliy Sen'ko*, seen here at Engels-2, is the latest Tu-160 to be christened. Note the cylindrical hut for the ground personnel. The mounds in the background are storage bunkers, not earthen revetments.

'11 Red' is named after the only Long-Range Aviation navigator to become twice Hero of the Soviet Union during the Second World War, hence the two Orders of the Gold Star painted on the fuselage (this order went with the HSU title).

Above: The all-movable upper fin of the Tu-160 is often left at maximum deflection on the ground, giving the uncanny impression that the aircraft is damaged and withdrawn from use – an impression enhanced by the rather grubby appearance of '33 Red' seen here stored at Priluki.

The christening ceremony of Tu-160 '11 Red' *Vasiliy Sen'ko* at Engels-2 AB, with the appropriate brass band. Note that the Orders of the Gold Star are painted on the starboard side only (see previous page).

Above: The Tu-160 flight line at Engels-2 AB, with the two eagle-bedecked examples named *Il'ya Muromets* ('05 Red' and '06 Red') nearest to the camera.

Another view of '11 Red' *Vasiliy Sen'ko*, resplendent in the newly-adopted colours.

Above and below: The Tupolev OKB's demonstrator aircraft, '63 Grey' (c/n 84704217), in the static park at the MAKS-93 airshow at Zhukovskiy, surrounded by an Avia Akkord twin-engined light aircraft, a Lyavin Del'fin-2 homebuilt aircraft and a brand-new Tu-154M airliner. Some of LII's test aircraft are visible beyond.

fictitious 'documents giving export clearance' from the crews. It took a special official telegram from the State Customs Committee of Russia to allow the first two of the now-Russian aircraft to depart from Priluki and Uzin.

On 6th November 1999 Tu-160 '10 Red' took off from Priluki AB, landing at Engels-2 AB several hours later. The ferry flight was performed by Lt. Col. Aleksey Serebryakov (crew captain), Maj. Aleksey Kalinin (co-pilot), Lt. Col. Igor' Sazonov (navigator) and Maj. Yuriy Paltoosov (WSO).

As might be imagined, the aircraft were given a grand reception in Engels. When the bombers bearing Ukrainian roundels and shield-and-trident markings (or grey pentagons where the Soviet red stars had been painted out) had been lined up on the hardstand, Russian flags were ceremonially raised over them. DA Commander Lt. Gen. Mikhail Oparin and the Air Force's Chief of Combat Training Arkadiy Barsookov presented the crews which had ferried the bombers with valuable gifts.

Thus began the transfer of the Ukrainian *Blackjacks* to the 1096th GvTBAP which, until then, had six Tu-160s, five of which were fully operational. The unit later changed its identity to the 121st GvTBAP while remaining within the 22nd GvTBAD of the 37th Air Army. By the end of January 2000 all eleven bombers covered by the agreement had been flown to their new home in Engels.

The last aircraft to arrive was also a Tu-160, an example coded '18 Red'. The aircraft had not flown for nearly nine years and was in pretty bad shape; it took a maximum of effort from the Russian tech staff on temporary duty at Priluki AB to return the bomber to airworthy status. All spare parts to replace defective or time-expired units had to be delivered from Russia.

Five of the eight *Blackjacks* were flown to Engels by a 121st GvTBAP crew captained by Aleksey Serebryakov; a crew under Lt. Col. Igor' Skitskiy ferried the remaining three. The cruise missiles were delivered to Russia by rail. Now the Russian Air Force had fifteen Tu-160s – enough to equip two full squadrons (and a second *Blackjack* squadron was indeed formed in due course). The total worth of the bombers was about US$ 285 million.

According to the now-former Long-Range Aviation (37th VA) Commander Mikhail Oparin, all the aircraft transferred from the Ukraine were in good condition (presumably this means 'after all defective or missing units had been replaced' – *Auth*.). The engines had used up only about 10% of their service life and would not need to be replaced for more than 20 years.

All remaining Tu-160s in the Ukraine were to be destroyed – though, as mentioned earlier, there were plans to sell two of them to the USA for civil use. One of the *Blackjacks* has been preserved for posterity at the Ukrainian Air Force Museum in Poltava.

Even before the second Tu-160 squadron was established, the 37th VA resumed intensive combat training in keeping with personal instructions from the President – a decision expedited, no doubt, by NATO's intervention in ex-Yugoslavia (Operation *Allied Force*). The war in ex-Yugoslavia was primarily an air war – something the decision makers in Russia could not miss.

Thus the spring of 1998 saw yet another command and staff exercise of the Russian Armed Forces. In the course of the exercise two *Blackjacks* from Engels flew an ultra-long-range sortie to the North Pole. A third Tu-160 captained by Guards Lt. Col. Aleksey Serebryakov accurately launched a Kh-55SM cruise missile against a practice target.

Another major exercise called ***Zahpad-99*** (West-99) took place in the summer of 1999. On the night of 26th June a pair of Tu-160s and a pair of Tu-95MSs departed from Engels-2 AB, heading north. The *Blackjacks* were captained by Guards Lt. Col. Igor' Skitskiy and Guards Lt. Col. Vladimir Popov. After reaching a predesignated point over the Polar Sea they turned south-west, parting company near the Norwegian coast. The two aircraft passed along the entire Norwegian coastline,

Another view of Tu-160 '63 Grey' at MAKS-93, with the Tu-134 IMARK geophysical research aircraft and a Tu-142MZ Bear-F Mod 3 ASW aircraft visible beyond.

Above: '01 Red' *Mikhail Gromov* **makes a low-speed flypast with 'everything down' for the benefit of the cameraman.**

Tu-160 '02 Red' named after Vasiliy Reshetnikov (the Long-Range Aviation Commander of the 1970s) comes in to land. Note the angle of the horizontal tail.

Above: '05 Red' basks in the sun at Engels. The green metal boxes on the apron are for the cables of the centralised power supply system, obviating the need for mobile power carts.

It is standard operational procedure to leave the wings at minimum sweep when on the ground. True, this increases the required apron space but it eliminates the danger of tipping the aircraft over on its tail – and then it is quite hard to return the aircraft to normal position without causing further damage.

Above: '03 Red' races down the runway at Engels, about to take off on a late afternoon sortie.

A 121st GvTBAP *Blackjack* crew heads for the debriefing room after a mission; the camouflaged flying suits are noteworthy. The vehicle in the background is a TZ-22, the most common Soviet airfield fuel bowser consisting of a KrAZ-257 tractor with a 22,000-litre (4,840-Imp. gal.) tank semi-trailer.

Above: Tu-160 '03 Red' taxies at Engels.

This full frontal of '12 Red' *Aleksandr Novikov* shows to advantage the shape of the flight deck roof and the narrow landing gear track.

109

Above: Lit by the failing evening light filtering through clouds, a Tu-160 makes a banking turn over lakeland.

Another air-to-air of '03 Red' which emphasises the Tu-160's slender and sharp-nosed profile.

making simulated missile launches. Then one of the crews proceeded to a target range in southern Russia where a Kh-55SM missile was launched for real. The Tu-160s stayed airborne for 12 hours on that occasion – without in-flight refuelling.

The Western media immediately raised Cain, claiming that 'Russia was planning a nuclear attack on the USA' or that 'Moscow had intruded into Icelandic airspace'. Both claims were in fact false – the Russian Air Force's 37th Air Army had simply given another effective demonstration of its capabilities.

Yet another show of force in which the Tu-160 was involved took place on 18th-21st April 2000 when the Russian Air Force held what was officially called a 'flight and tactical exercise'. As the Russian military put it, 'you have to show force at the right time in order to avoid actually using force later'. *Si vis pacem, para bellum*. The main purpose of the exercise was to 'check the combat readiness of the aviation hardware after a prolonged grounding and improve the participants' proficiency'. This time the ex-Ukrainian Tu-160s joined the fun, launching cruise missiles together with *Bear-Hs*, while Tu-22M3 bombers attacked their targets with free-fall bombs. Interaction with ground command, control and communications (C^3) centres and targeting with the help of Il'yushin/Beriyev A-50 *Mainstay-A/B* airborne warning and control system (AWACS) aircraft during air defence penetration in an ECM environment were also practiced.

The 2000 exercise featured an important 'first' – for the first time the Russian Air Force practiced using precision-guided cruise missiles with conventional warheads. This again was a result of analysing the course of the Gulf War (Operation *Desert Storm*) and the war in ex-Yugoslavia where the USA had used Tomahawk cruise missiles on a large scale.

Together with the formation of Russian's first Tu-160 squadron a long-forgotten tradition was reborn. Nose art was generally frowned upon in the Soviet Air Force (except in times of war when it helped keep up the fighting spirit), being broadly regarded as characteristic of the potential enemy and hence unbecoming for the Soviet aces. Now, however, the majority of the Tu-160 fleet received proper names (see table below). Curiously, the first two aircraft to be christened, '05 Red' and '06 Red', bore the same name – *Il'ya Muromets* (a Russian epic hero) – and rather non-standard markings in the form of blue/yellow nose flash (a fragment of the Air Force's 'sunburst' flag), a Russian flag tail flash and the Russian 'double eagle' on the tail in lieu of the red star. (The latter was probably at a time when the Russian Air Force was experimenting with various new insignia before eventually deciding to leave well enough alone. After all, the Red Star is immediately recognisable, unlike assorted roundels which often give rise to confusion.)

On one photo the two aircraft were pictured together at Engels, and in 1999 the editorial staff of the *Avi**ah**tsiya i Kosmo**nahv**tika* (Aviation and Spaceflight) monthly 'doctored' the picture a little, changing the name of '05 Red' to *Il'yin Muromets*! This was a practical joke aimed at aviation writer Vladimir Il'yin who was a member of the magazine's editorial board.

On the aircraft christened later the colour scheme was changed a little. Now most 121st GvTBAP Tu-160s do not have the 'double eagle'; instead, they combine a narrow Russian flag fin flash along the fin leading edge with the red star.

Tu-160 operational details

Tactical code	C/n	Unit	Notes
01 Red		121st GvTBAP, Engels	Named 'Mikhail Gromov' after a famous test pilot
02 Red		121st GvTBAP, Engels	Named 'Vasiliy Reshetnikov' after the Long-Range Aviation Commander during the 1970s
03 Red		121st GvTBAP, Engels	
04 Red		121st GvTBAP, Engels	Named 'Ivan Yarygin' after a famous Soviet wrestler who won the Olympic gold medal in 1972
05 Red		121st GvTBAP, Engels	Named 'Il'ya Muromets' (the first to be thus named); later rechristened 'Aleksandr Golovanov' after the Long-Range Aviation Commander in 1942-44 and 1946-48
06 Red		121st GvTBAP, Engels	Named 'Il'ya Muromets' (the second to be thus named)
07 Red	? (f/n 0802)	121st GvTBAP, Engels	D/D 3-5-00; named 'Aleksandr Molodchiy' after a famous bomber pilot and Hero of the Soviet Union
10 Red		184th GvTBAP, Priluki	To Ukraine AF; sold to Russia in 1999
		121st GvTBAP, Engels	
11 Red		184th GvTBAP, Priluki	To Ukraine AF; sold to Russia in 1999
		121st GvTBAP, Engels	Named 'Vasiliy Sen'ko' after the sole Long-Range Aviation navigator to become twice Hero of the Soviet Union
12 Red		184th GvTBAP, Priluki	Displayed at Kubinka AB 2-8-88 to Frank Carlucci; sold to Russia in 1999
		121st GvTBAP, Engels	Named 'Aleksandr Novikov' after another Soviet Air Chief Marshal of the 1940s
14 Red		184th GvTBAP, Priluki	Mfd 1991. To Ukraine AF. Scrapped 11-99 as the second Tu-160 to be destroyed
15 Red		184th GvTBAP, Priluki	
16 Red		184th GvTBAP, Priluki	Displayed at Machoolischchi AB 13-2-92; sold to Russia in 1999. Named 'Aleksey Plokhov' in 2002 after a bomber pilot and Hero of the Soviet Union
17 Red		184th GvTBAP, Priluki	
18 Red		184th GvTBAP, Priluki	
20 Red		184th GvTBAP, Priluki	
21 Red		184th GvTBAP, Priluki	Displayed at Kubinka AB 13-7-89 to Adm. William Crowe
24 Red		184th GvTBAP, Priluki	Mfd 1989. To Ukraine AF; scrapped 16-11-98 as the first Tu-160 to be destroyed, total time 466 hrs
25 Red		184th GvTBAP, Priluki	To Ukraine AF
26 Red		184th GvTBAP, Priluki	To Ukraine AF
29 Grey		Tupolev OKB	Stored Zhukovskiy
33 Red		184th GvTBAP, Priluki	To Ukraine AF
63 Grey	84704217	Tupolev OKB	Stored Zhukovskiy
86 Grey		Tupolev OKB	Stored Zhukovskiy
87 Grey		Tupolev OKB	Stored Zhukovskiy
not known		121st GvTBAP, Engels	Named 'Pavel Taran' in 2002 after a bomber pilot

This magnificent plan view of a *Blackjack* with the wings at intermediate sweep (35°) illustrates well why the Tu-160 is called *Belyy Lebed'* (White Swan).

Chapter 6

Blackjack vs 'Bone'

Equals or Not?

Some might be tempted to put the question differently: copy or not? Sure enough, the Tu-160 and the Rockwell International B-1 look quite similar at first glance. Much has been said about the apparent Soviet custom of copying Western designs; this postulate is rooted in a firm conviction that Russia, the old Cold War enemy, cannot produce anything worthwhile. However, it is no surprise that the engineers developing both aircraft chose the same general arrangement, aerodynamic features and internal layout more than 20 years ago. Ideas are borne on the wind, as a Russian saying goes; and indeed, faced with similar general requirements and given basically the same levels of aviation science and technology, the two nations were bound to come up with similar results. Yet, a closer look at the two bombers reveals that the Tu-160 and the B-1 are not so similar after all.

Born under President Richard Nixon, the B-1 had a head start on the *Blackjack*; the first prototype of the original B-1A (USAF serial 74-0158) first flew on 23rd December 1974, followed by three other prototypes, one of which was originally a structural test airframe. However, mounting programme costs worried the new President, Jimmy Carter (known for his 'belt-tightening' policies) so much that he finally cancelled the B-1 on 30th June 1977, the last day of Fiscal Year 1977. Yet the subsequent revelation that the Soviet Union was working on a new strategic bomber programme prompted the US Department of Defense to revive the programme, adapting it to changed priorities; the result was the B-1B Lancer – or, as it is popularly known, the Bone (a corruption of 'B-One').

During the transformation from A to B Rockwell spent a lot of effort on reducing the bomber's radar cross-section; a new, more fuel-efficient version of the General Electric F101 afterburning turbofan was fitted, and the avionics and armament were revised. As a result, the maximum take-off weight rose from the B-1A's 180 tons (395,000 lb) to 217 tons (477,000 lb). However, the B-1 lobby and the US Air Force did not succeed in proving to the US Congress the need to incorporate a whole range of costly features into the B-1B's design and the Congress slashed the funds for the new bomber. Consequently the engineers had to use rather less titanium than they wanted to and use simple fixed-area air intakes instead of variable supersonic intakes; the latter restricted the bomber's top speed to Mach 1.25. The armament was to consist of Boeing AGM-86B (ALCM) cruise missiles, Lockheed AGM-69A (SRAM) short-range attack missiles and nuclear bombs.

The B-1B prototype (82-0001 *Leader of the Fleet*, c/n 1) entered flight test on 23rd March 1983; it remained a test aircraft and was never delivered to the USAF. The first production aircraft (83-0065 *Star of Abilene*, c/n 2) took off on 18th October 1984. The 100th and final B-1B (86-0140 *Valda J*, c/n 100) left the production line in Palmdale, California, in 1988.

Conversely, the Tu-160 was developed by the world's second superpower at a time when funding issues were of minor importance, if any – in those days the Soviet military got all the money they wanted, as long as the required weapons systems were developed and fielded on time. Hence the Tu-160 escaped the 'vivisection' the B-1 had been subjected to, and the aircraft which entered production and service with the Soviet Air Force was exactly what its creators had wanted it to be – a multi-mode aircraft capable of delivering intercontinental strikes within a wide altitude and speed envelope.

On the other hand, the production line at Palmdale was turning out a steady stream of Lancers on (or ahead of) schedule and the B-1B was already fully operational at a time when Tu-160 production in Kazan' was only just commencing. Today the 'Bone', together with the long-serving Boeing B-52H Superfortress and a small number of the highly sophisticated Northrop B-2A Spirit stealth bombers, makes up the backbone of the USAF's strategic component.

After the demise of the Soviet Union the balance of power shifted; Russia had to work hard in order to at least partly rebuild it strategic bomber force. Despite these efforts, today Russia has only a single regiment of Tu-160s – sixteen aircraft, which is equivalent to just over 15% of the USAF's B-1 fleet.

As for the capabilities of the two aircraft, they can be compared only in theory.

Sure, outwardly the Tu-160 is very similar to the B-1B, having the same general arrangement, utilising the same blended wing/body layout and variable-geometry wing design. However, the Russian bomber is much larger and heavier, which is why the aggregate thrust of its engines is 79% higher. The operating speeds are quite different as well. As already noted, at the insistence of the USAF Rockwell had to do without variable supersonic air intakes. Hence at high altitude the B-1B cannot exceed Mach 1.2, which is not ideal from a tactical standpoint. Conversely, the Tu-160 can reach 2,200 km/h (1,366 mph; 1,189 kts), thanks to its variable intakes, ample engine thrust and slender fuselage having a relatively small cross-section area. Low drag was attained thanks not only to streamlined contours but also to a carefully designed internal layout thanks to which the Tu-160's fuselage height is no bigger than that of the much smaller Tu-22M3.

Also, the Tu-160 is designed in such a way as to achieve maximum possible range both in high-altitude supersonic cruise and in ultra-low-level flight. These modes can be used separately or in a combination to fufil the mission with maximum efficiency. This is the Russian bomber's multi-mode design philosophy.

The *Blackjack* has an advantage in offensive capability as well – its main weapon, the Kh-55SM cruise missile, is well mastered by both the industry and the bomber crews. Conversely, the Americans were unable to adapt the B-1B to the costly AGM-86B due to budgetary constraints – this would require not only the bomb bays to be modified but also the avionics suite to be substantially altered. The AGM-69A had to be excluded from the inventory in 1994 because the stockpile of missiles had reached the end of their shelf life and the solid propellant had started decomposing. This left the B-1B with only the B61 and B83 free-fall nuclear bombs (though a small number of B28 nukes remained available in 1996). (Note: As of 1996 the USAF had plans to integrate the General Dynamics AGM-129A (ACM) advanced cruise missile on the B-1B. The Boeing AGM-131A (SRAM II) was also proposed but was cancelled in September 1991.)

As for conventional munitions, the Lancer did not receive a conventional capability until after the Gulf War (true, live weapons tests began in 1991 but the fleet-wide upgrade

Above: The photos on these pages make an interesting comparison. Here, 1096th GvTBAP Tu-160s are seen on the flightline at Engels-2 AB in the days when '05 Red' was still named *Il'ya Muromets*.

Another point for comparison. Seen by the refuelling systems operator of an IL-78, a Tu-160 closes in on the tanker with the refuelling probe extended. The bulges on the wing upper surface just inboard of the engine nacelles are fairings for the main gear bogies' outer rows of wheels.

Above: A line-up of 96th Bombardment Wing B-1Bs at Dyess AFB, Texas, in 1991. Unlike the Tu-160s which have an all-white 'anti-nuclear' finish, the Lancers invariably wear a dark two-tone camouflage scheme. Another notable difference is the provision of an integral boarding ladder which the Tu-160 lacks.

A similar perspective of a B-1B (85-0072, c/n 32) formating with a tanker clearly shows the characteristic Low-Altitude Ride Control (LARC) vanes (now called SMCS – Structural Mode Control System), the aerial refuelling receptacle, the larger flight deck glazing area and the generally leaner contours of the aircraft.

Above: In some circumstances the *Blackjack*'s overall white colour scheme can be an effective camouflage in its own right, as illustrated by the as-yet unnamed '06 Red' cruising with the wings at 35°. This view also illustrates the Tu-160's sleek profile accounted for by the high fineness ratio of the fuselage.

came too late for the action); it was first used operationally during the war in ex-Yugoslavia (Operation *Allied Force*). Conversely, the Tu-160 had a conventional capability from the start, hence the inclusion of the OPB-15T electro-optical bombsight into the targeting suite.

The approach to weapons carriage is different, too. The B-1B has three weapons bays (two ahead of the wing pivot box carry-through and one aft), while the Tu-160 has two bays of larger dimensions. Also, the Lancer has provisions for carrying missiles on six external hardpoints under the forward, centre and rear fuselage, whereas on the *Blackjack* all armament is stowed internally. This helps reduce the bomber's RCS and reduce drag, thereby increasing range – albeit this also accounts for the larger size of the Tu-160.

In avionics and equipment, the B-1B apparently comes out on top. According to press reports, Russian and Ukrainian pilots described the Lancer's flight instrumentation as 'excellent'. As regards crew comfort and cockpit ergonomics, the two aircraft are about equal, although the B-1B's flight deck offers somewhat less headroom, being encroached on from below by the nosewheel well. As for the mission avionics, some Russian systems are theoretically more capable than their US counterparts but are not used in full or not used at all for various reasons (reliability problems etc.); also, some of the *Blackjack*'s avionics are still hampered by operational limits imposed in some flight modes.

The B-1 (illustrated here by B-1A 74-0160 following modifications under the B-1B development programme and repainted in a desert camouflage scheme) has a distinctive humpbacked appearance. The spine housing ECM equipment was omitted on the B-1B as the system was stowed inside the fuselage.

The Russian military and many of the world's top aviation experts believe that the combination of the Tu-160's performance characteristics and design features theoretically gives it an edge over the B-1B and other American bombers, including the stealthy B-2A – but theory is one thing and real life is another. Due to persistent funding shortfalls the Russian Air Force is currently unable to maintain its operational bomber fleet in perfect condition – and apparently will not be able to do so in the foreseeable future (to say nothing of providing enough flying hours for the crews). Maintaining proficiency is a sore problem for the Russian airmen. For instance, both the 'Bone' and the *Blackjack* have IFR capability; however, B-1B pilots practice aerial refuelling almost weekly – something their Russian colleagues can only dream of.

An opportunity to make an objective comparison of the two types came on 23rd-25th September 1994 when the Tu-160 and the B-1B 'rubbed noses' (fortunately not literally) for the first time at Poltava AB during the Shuttle Raid 50th anniversary celebrations, to which the USAF sent a large delegation. The flight and ground crews of both bombers had a chance to examine each other's aircraft and form an opinion for themselves.

Here is the opinion of 37th VA Commander Lt. Gen. Mikhail Oparin:

'I have a deep respect for the people who charted the development perspectives for the Long-Range Aviation in the 1980s/early 1990s time frame. The structural strength reserves and upgrade potential of the Tu-95MS and Tu-160 strategic bombers allows them to be called aircraft of the 21st century, and with good reason – the missile-carrying aircraft still have unused potential. These bombers are not only a match for the best Western hardware but excel it in certain respects. I know what I'm saying because I have had a chance to study the strategic aircraft of our 'friends and rivals' firsthand. I had the opportunity to fly a real B-52, and I made several flights in the B-1 simulator; after this I was enchanted by the Tu-95MS and especially the Tu-160.'

It would be best to conclude this chapter, and the book at large, with the following words said by former Russian Air Force C-in-C Army General Pyotr S. Deynekin:

'What makes the best comparison for the "Il'ya Muromets" (ie, the Tu-160 – *Auth*.)? The Tu-95? Or perhaps the [Antonov] An-124 (the world's heaviest operational transport aircraft – *Auth*.)? I guess the correct answer is the B-1B Lancer, the Tu-160's American counterpart. In May 1992 I made three flights in a B-1B over the Nevada Desert, flying the bomber from the left-hand seat, with multiple top-ups from a KC-135 tanker. I have a commemorative picture signed by the Commanders of USAF bomber wings. I daresay they are both good aircraft and worthy rivals – as, incidentally, are the men who fly them. This is why we'd better be friends than foes, and the Americans are well aware of this.'

Specifications of Soviet and US strategic bombers

	Northrop B-2A	Rockwell B-1B	Boeing B-52H	Tu-160	Tu-95MS
Crew	2	4	6	4	7
Powerplant	4 x General Electric F118-GE-100	4 x General Electric F101-GE-102	8 x Pratt & Whitney TF33-PW-3	4 x Kuznetsov NK-32	4 x Kuznetsov NK-12MP w. AV-90 props
Max power, kgp (lb st)	4 x 8,600 (4 x 18,960)	4 x 10,610 (4 x 23,390)	8 x 7,700 (8 x 16,975)	4 x 25,000 (4 x 55,115)	–
Max power, ehp	–	–	–	–	4 x 15,000
Length overall	21.0 m (68 ft 10¾ in)	45.78 m (150 ft 2½ in)	47.8 m (156 ft 10 in)	54.095 m (177 ft 5¾ in)	49.13 m (161 ft 1 in) *
Height on ground	5.45 m (17 ft 10½ in)	10.24 m (33 ft 7¾ in)	12.41 m (40 ft 8½ in)	13.0 m (42 ft 8 in)	13.3 m (43 ft 7½ in)
Wing span	52.4 m (171 ft 11 in)	41.67/23.84 m (136 ft 8½ in/ 78 ft 2½ in) **	56.39 m (185 ft)	57.7/35.6 m (189 ft 3¾ in/ 116 ft 9½ in) **	50.04 m (164 ft 2 in)
Wing area, m² (sq ft)	465.0 (5,000)	181.2 (1,950)	371.6 (3,995)	293.15 (3,152.15)	289.9 (3,117)
MTOW, kg (lb)	181,500 (400,130)	216,400 (477,000)	221,400 (488,100)	275,000 (606,260)	185,000 (407,850)
Ordnance, kg (lb)	22,600 (49,820)	34,000 (75,000)	23,000 (50,700)	40,000 (88,180)	7,800/20,800 (17,195/ 45,855) ***
Fuel load, kg (lb)	73,000 (160,930)	88,450 (195,000)	181,800 (400,800)	171,000 (376,980)	87,000 (191,799)
Speed, km/h (mph):					
cruising	850 (528)	855 (531)	837 (520)	1,000 (621)	750 (465)
maximum	1,000 (621)	1,270 (789)	1,014 (630)	2,000 (1,242)	830 (515)
Service ceiling, m (ft)	12,500 (41,000)	15,240 (50,000)	16,700 (54,790)	15,000 (49,210)	10,500 (34,440)
Range, loaded, km (miles):					
on internal fuel	11,100 (6,890)	5,540 (3,444)	15,470 (9,600)	12,300 (7,640)	10,500 (6,521)
with one top-up	18,500 (11,490)	n.a.	n.a.	n.a.	14,100 (8,756)

* With IFR probe
** At minimum/maximum sweep
*** Tu-95MS-6/Tu-95MS-16 (with six or 16 Kh-55MS missiles respectively)

Upper view of a production Tu-160 with the wings at maximum sweep.

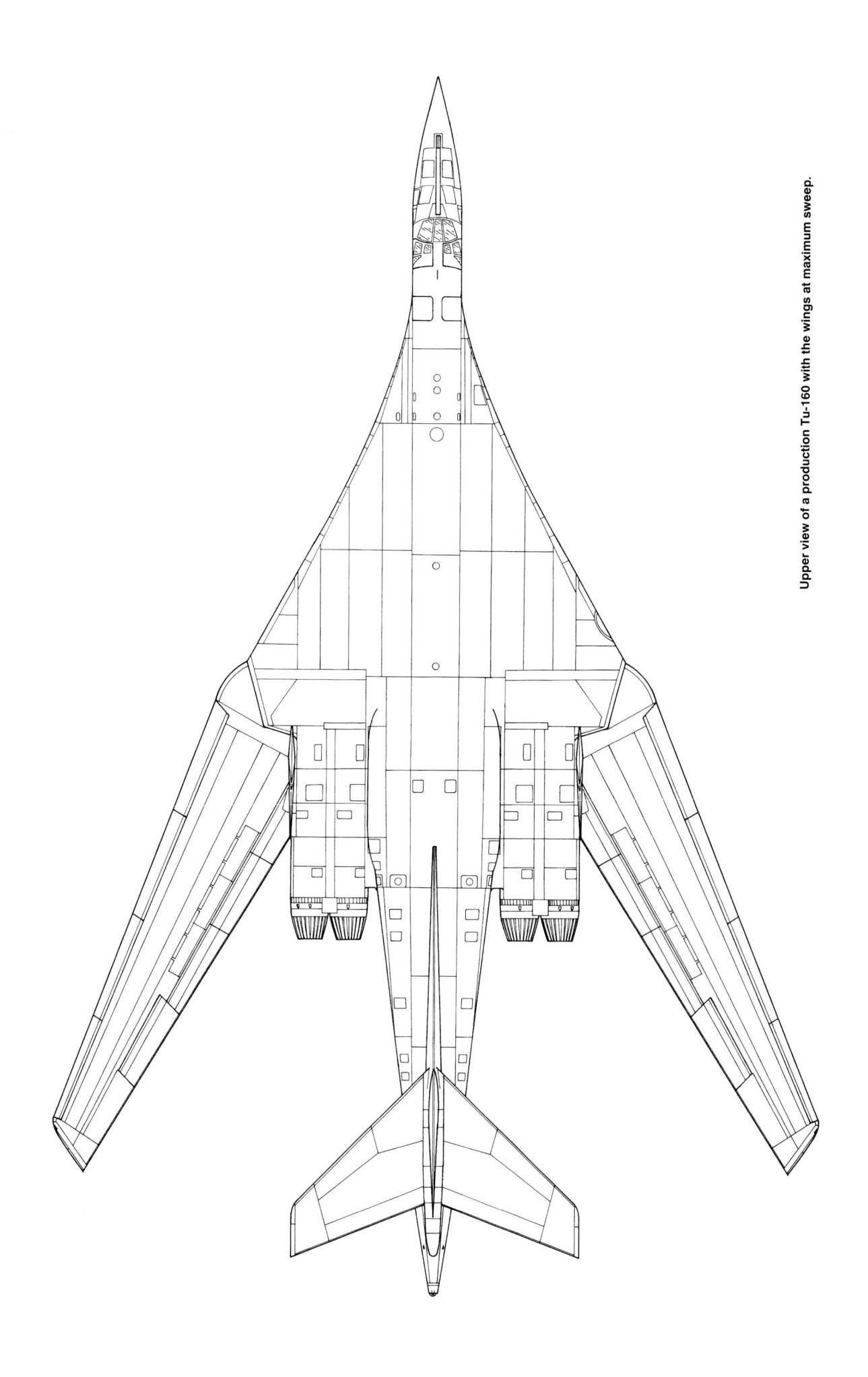

Lower view of a production Tu-160 with the wings at maximum sweep.

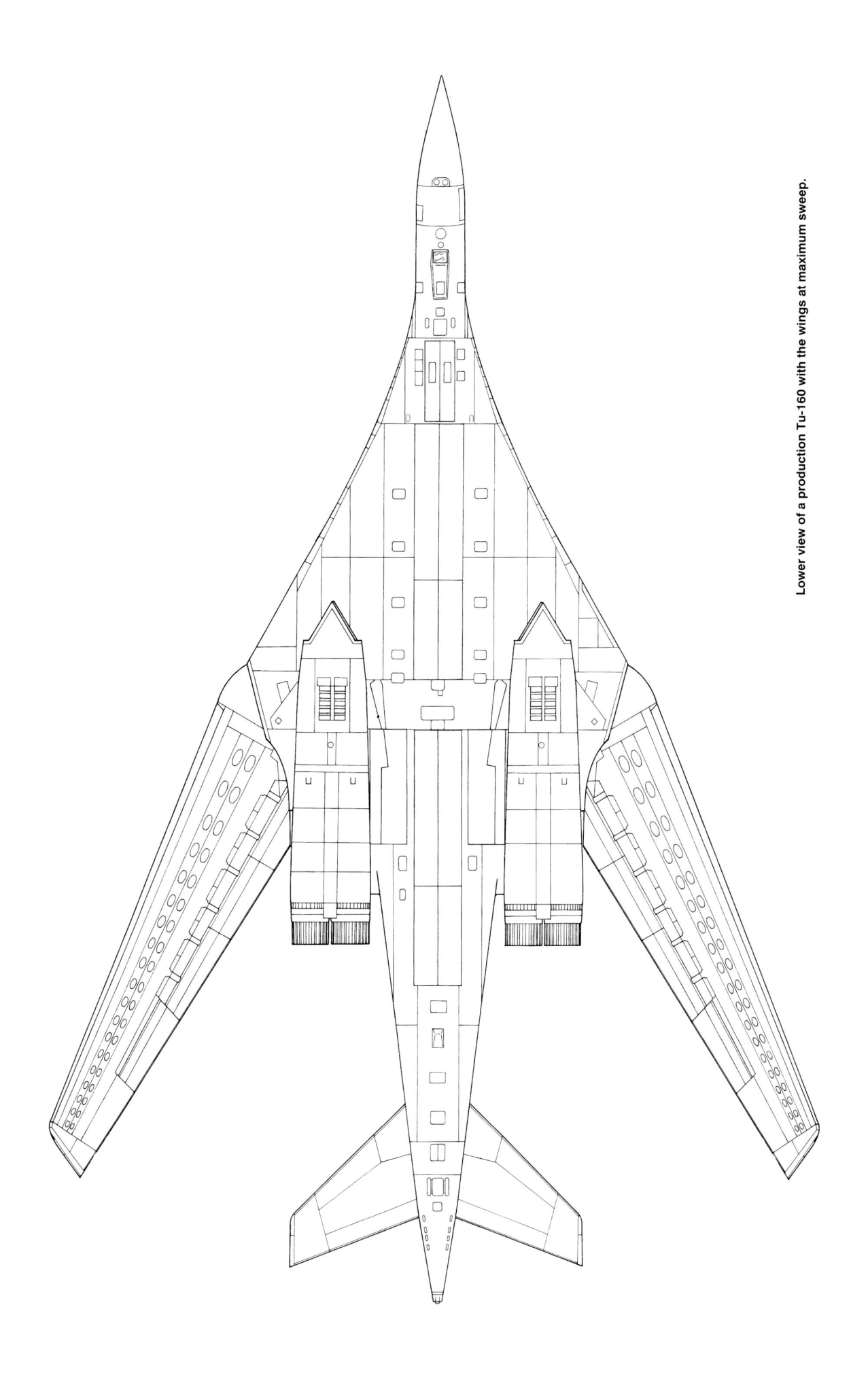

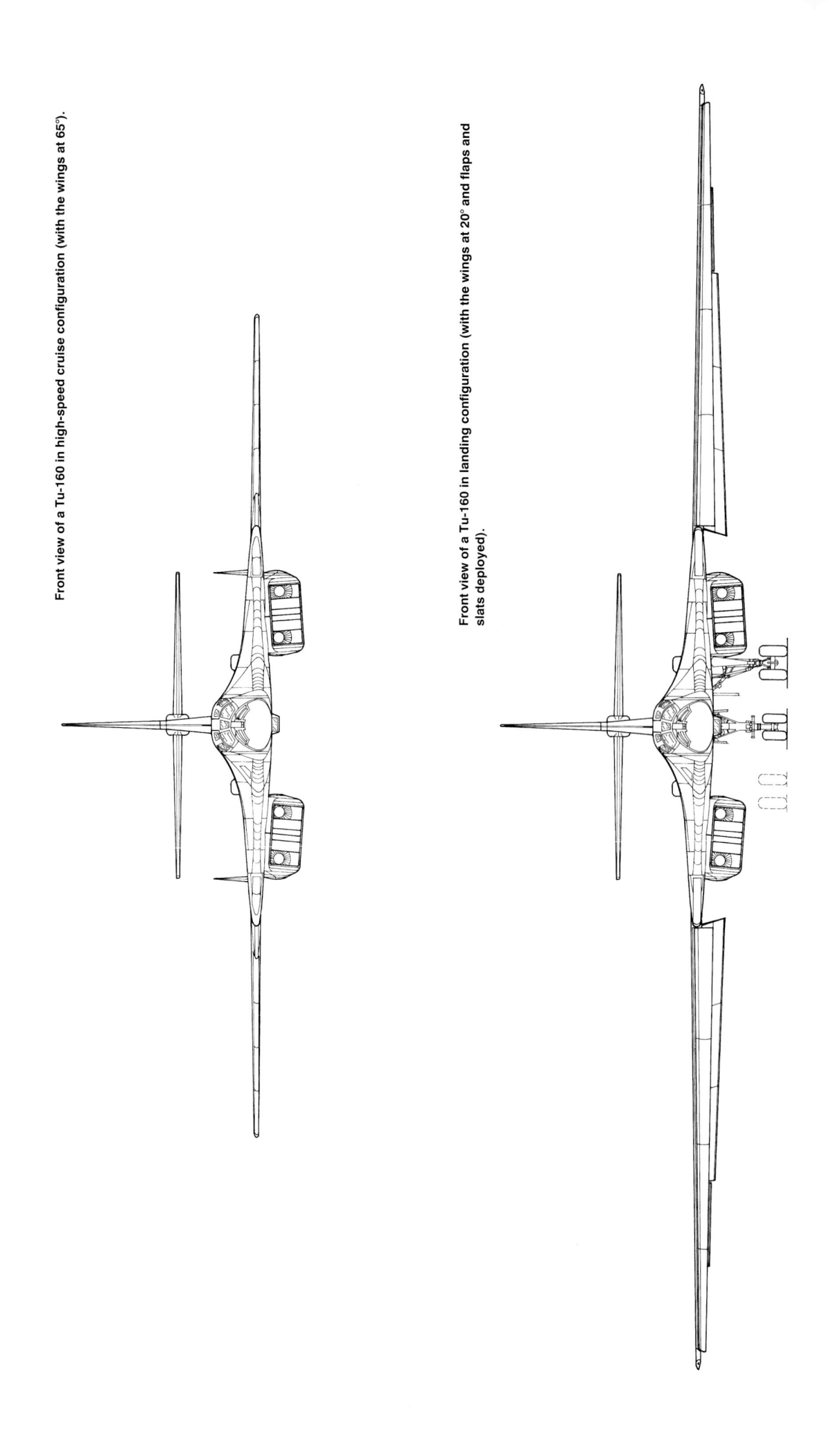

Front view of a Tu-160 in high-speed cruise configuration (with the wings at 65°).

Front view of a Tu-160 in landing configuration (with the wings at 20° and flaps and slats deployed).

Starboard side view of a production Tu-160 in high-speed cruise configuration.

Port side view of a production Tu-160 with the wings at 20° and flaps and slats deployed.

Below: This Tu-160 development aircraft never received a coat of paint. There were two more of these 'silver bullets'.

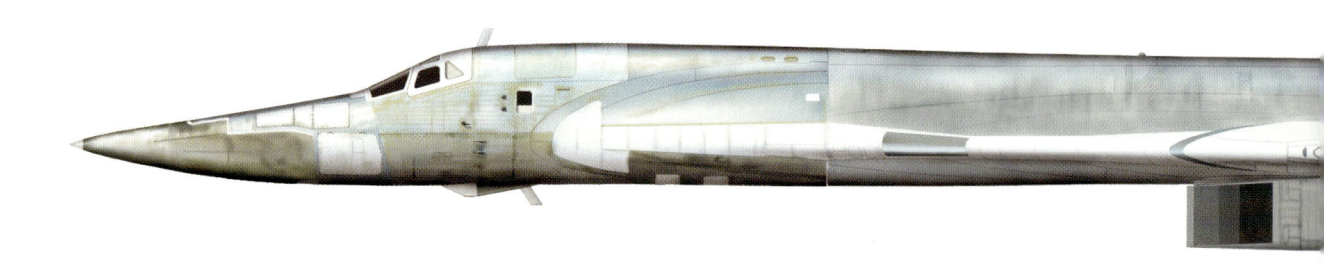

Below: Tu-160 '01 Red' *Mikhail Gromov* is one of several 37th VA/22nd GvTBAD/121st GvTBAP aircraft to have a Russian Air Force flag nose flash and a Russian flag fin flash and double eagle. Note the six mission markers denoting successful missile launches.

Below: '06 Red' *Il'ya Muromets*, one of the first two Tu-160s to be christened (at the time when the unit was still the 1096th GvTBAP).

123

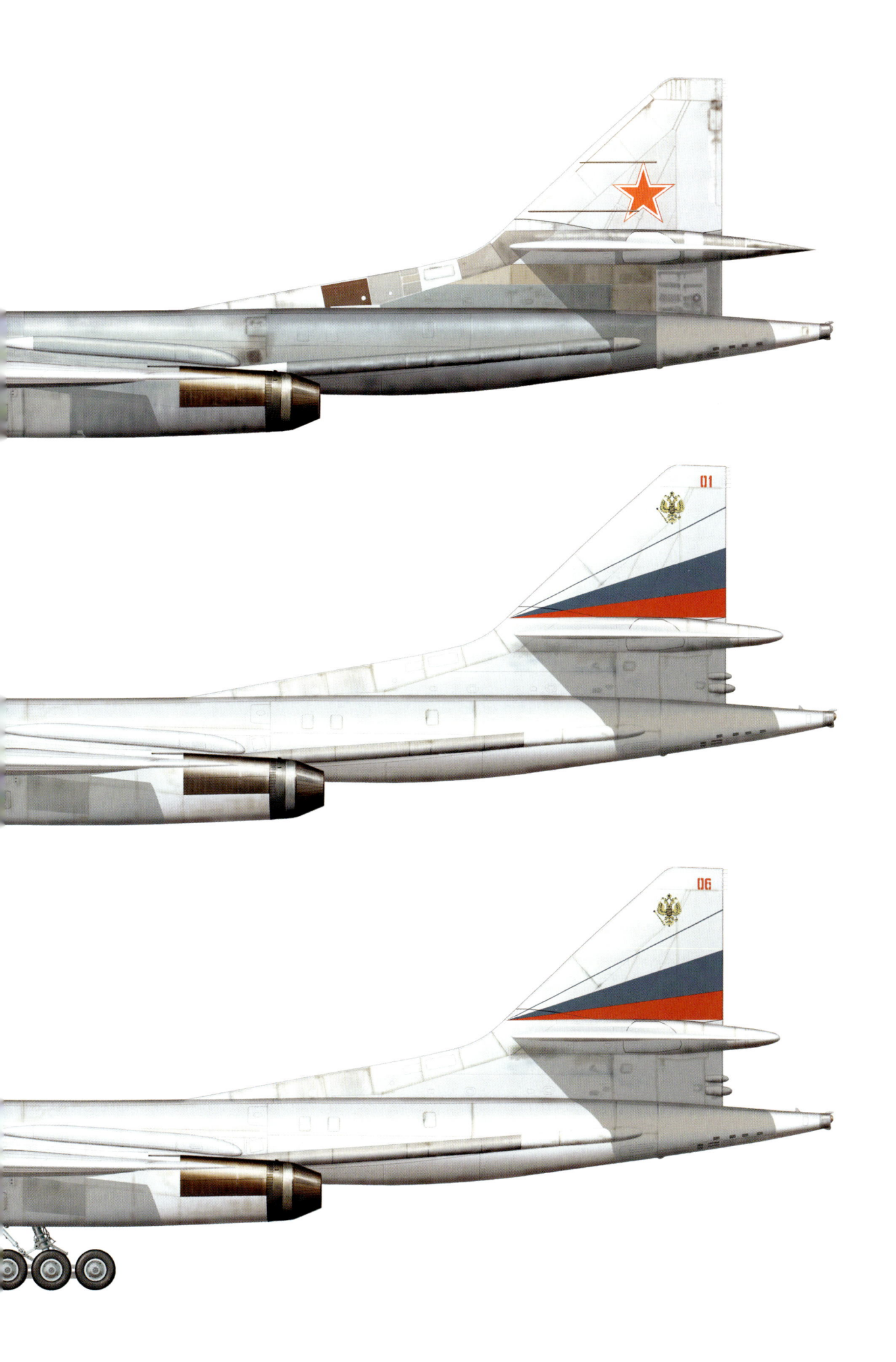

Below: 121st GvTBAP Tu-160 '02 Red' is named after former Long-Range Aviation Commander Colonel General Vasiliy V. Reshetnikov.

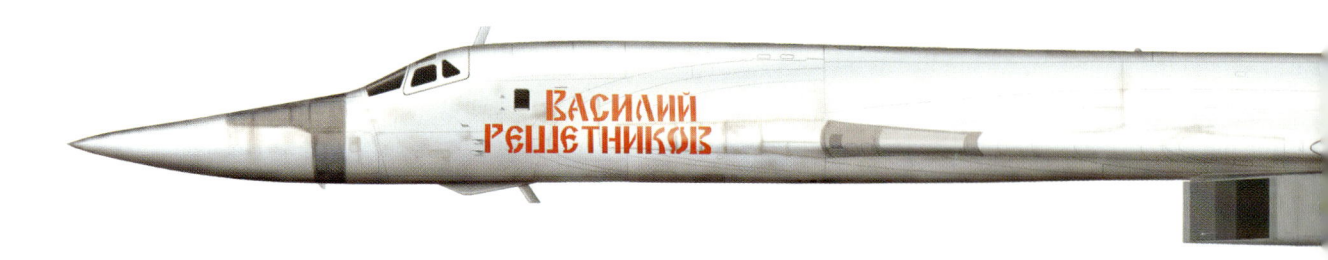

Below: '04 Red' is named after the Soviet wrestler Ivan Yarygin who won the gold at the 1972 Olympic Games in Munich. Note the three mission markers.

Below: '07 Red' (f/n 0802), the latest new-build *Blackjack* to be delivered to the Russian Air Force, is named *Aleksandr Molodchiy* after a famous bomber pilot and Hero of the Soviet Union. Note that the tactical code is not repeated on the tail on this aircraft.

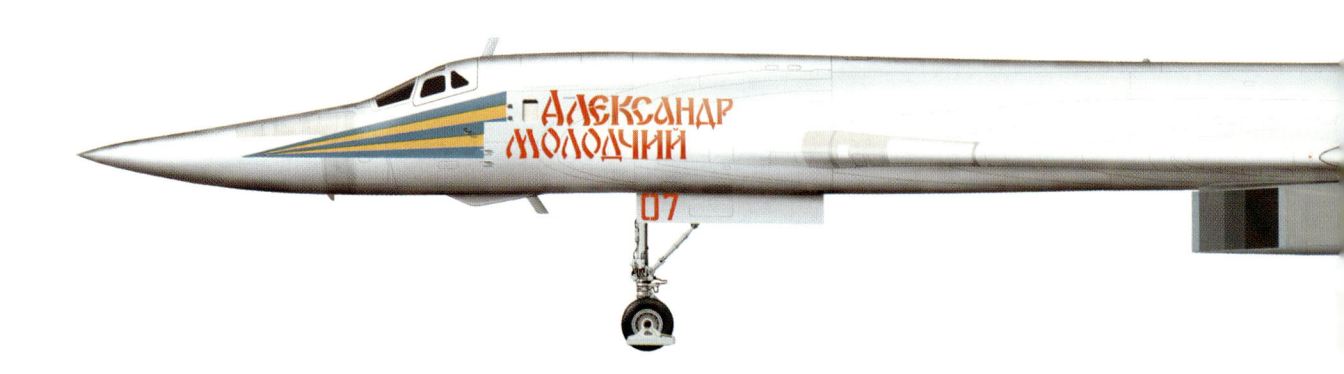

125

Below: Originally '05 Red' was named *Il'ya Muromets* until someone decided that having two aircraft of the same name was improper and '05 Red' was renamed after Air Marshal Aleksandr Golovanov, the Long-Range Aviation Commander in 1942-44 and 1946-48.

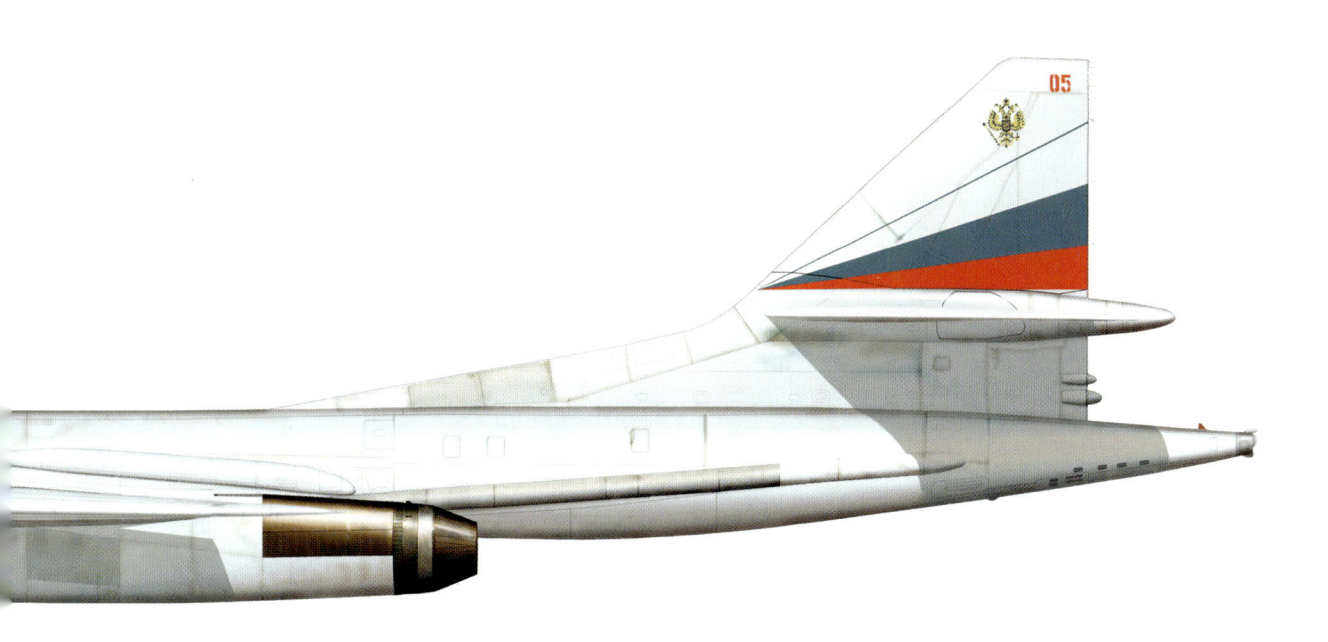

Below: Tu-160 '12 Red' is named *Aleksandr Novikov* after a Soviet Air Chief Marshal of the 1940s.

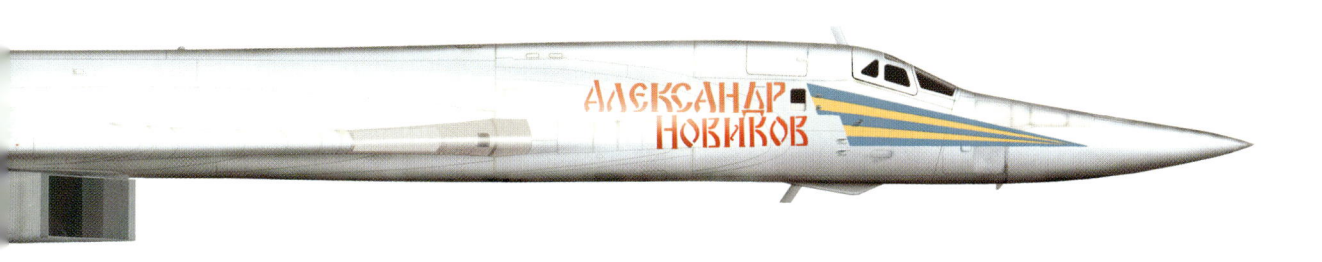

Below: '11 Red' is the latest Russian Air Force Tu-160 to be christened and is named after Vasiliy Sen'ko, the only Long-Range Aviation navigator to become twice Hero of the Soviet Union (as indicated by the two Gold Star orders on the fuselage).

We hope you enjoyed this book...

Midland Publishing titles are edited and designed by an experienced and enthusiastic team of specialists.

Further titles are in preparation but we always welcome ideas from authors or readers for books they would like to see published.

In addition, our associate, Midland Counties Publications, offers an exceptionally wide range of aviation, military, naval and transport books and videos for sale by mail-order worldwide.

For a copy of the appropriate catalogue, or to order further copies of this book, and any of many other Midland Publishing titles, please write, telephone, fax or e-mail to:

Midland Counties Publications
4 Watling Drive, Hinckley,
Leics, LE10 3EY, England

Tel: (+44) 01455 254 450
Fax: (+44) 01455 233 737
E-mail: midlandbooks@compuserve.com
www.midlandcountiessuperstore.com

US distribution by Specialty Press – see page 2.

Still available – Volume 1:
SUKHOI S-37 & MIKOYAN MFI
Yefim Gordon

Sbk, 280 x 215 mm, 96pp, 8pp foldout, 174 colour photos, dwgs, col artworks
1 85780 120 2   **£18.95/US $27.95**

Red Star Volume 2
FLANKERS: The New Generation
Yefim Gordon

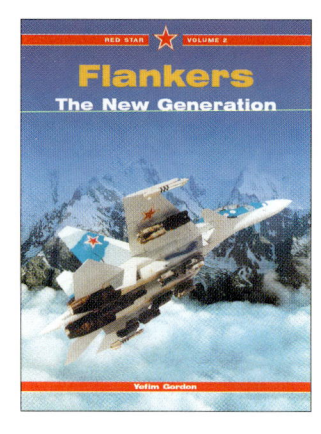

The multi-role Su-30 and Su-35 and thrust-vectoring Su-37 are described in detail, along with the 'big head' Su-23FN/Su-34 tactical bomber, the Su-27K (Su-33) shipborne fighter and its two-seat combat trainer derivative, the Su-27KUB. The book also describes the customised versions developed for foreign customers – the Su-30KI (Su-27KI), the Su-30MKI for India, the Su-30MKK for China and the latest Su-35UB.

Softback, 280 x 215 mm, 128 pages 252 colour photographs, plus 14 pages of colour artworks
1 85780 121 0   **£18.95/US $27.95**

Red Star Volume 3
POLIKARPOV'S I-16 FIGHTER
Yefim Gordon and Keith Dexter

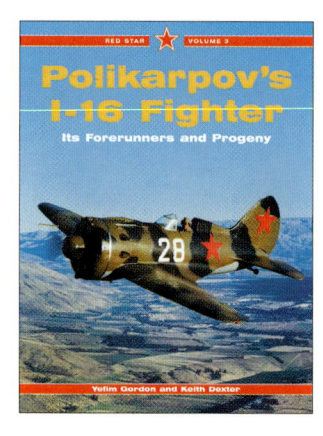

Often dismissed because it did not fare well against its more modern adversaries in the Second World War, Nikolay Polikarpov's I-16 was nevertheless an outstanding fighter – among other things, because it was the world's first monoplane fighter with a retractable undercarriage. Its capabilities were demonstrated effectively during the Spanish Civil War. Covers every variant, from development, unbuilt projects and the later designs that evolved from it.

Sbk, 280 x 215 mm, 128 pages, 185 b/w photographs, 17 pages of colour artworks, plus line drawings
1 85780 131 8   **£18.99/US $27.95**

Red Star Volume 4
EARLY SOVIET JET FIGHTERS
Yefim Gordon

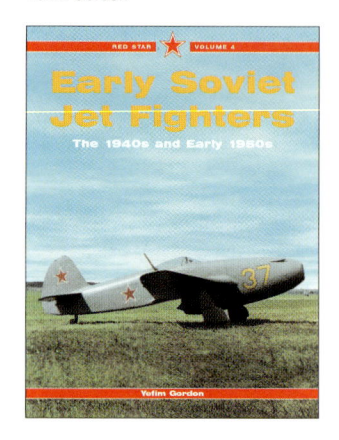

This charts the development and service history of the first-generation Soviet jet fighters designed by such renowned 'fighter makers' as Mikoyan, Yakovlev and Sukhoi, as well as design bureaux no longer in existence – the Lavochkin and Alekseyev OKBs, during the 1940s and early 1950s. Each type is detailed and compared to other contemporary jet fighters. As ever the extensive photo coverage includes much which is previously unseen.

Sbk, 280 x 215 mm, 144 pages 240 b/w and 9 colour photos, 8 pages of colour artworks
1 85780 139 3   **£19.99/US $29.95**

Red Star Volume 5
YAKOVLEV'S PISTON-ENGINED FIGHTERS
Yefim Gordon & Dmitriy Khazanov

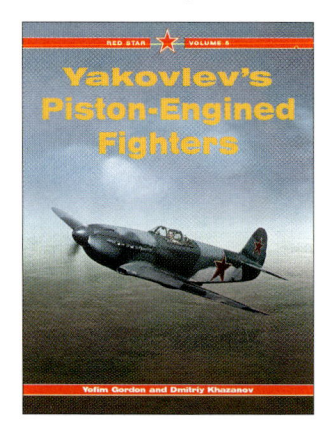

This authoritative monograph describes this entire family from the simple but rugged and agile Yak-1 through the Yak-7 (born as a trainer but eventually developed into a fighter) and the prolific and versatile Yak-9 to the most capable of the line, the Yak-3 with which even the aces of the Luftwaffe were reluctant to tangle. Yak piston fighters also served outside Russia and several examples can be seen in flying condition in the west.

Sbk, 280 x 215 mm, 144 pages, 313 b/w and 2 col photos, 7pp of colour artworks, 8pp of line drawings
1 85780 140 7   **£19.99/US $29.95**

Red Star Volume 6
POLIKARPOV'S BIPLANE FIGHTERS
Yefim Gordon and Keith Dexter

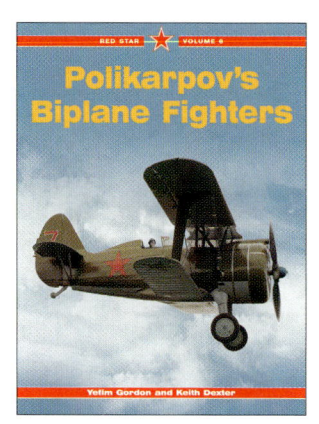

The development of Polikarpov's fighting biplanes including the 2I-N1, the I-3, and I-5, which paved the way for the I-15 which earned fame as the Chato during the Spanish Civil War and saw action against the Japanese; the I-15*bis* and the famous I-153 Chaika retractable gear gull-wing biplane. Details of combat use are given, plus structural descriptions, details of the ill-starred I-190, and of privately owned I-15*bis* and I-153s restored to fly.

Softback, 280 x 215 mm, 128 pages c250 b/w and colour photos; three-view drawings, 60+ colour side views
1 85780 141 5   **£18.99/US $27.95**

Red Star Volume 7
TUPOLEV Tu-4 SOVIET SUPERFORTRESS
Yefim Gordon and Vladimir Rigmant

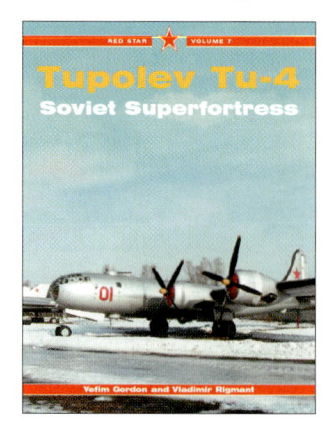

At the end of WW2, three Boeing B-29s fell into Soviet hands; from these came a Soviet copy of this famous bomber in the form of the Tu-4. This examines the evolution of the 'Superfortresski' and its further development into the Tu-70 transport. It also covers the civil airliner version, the Tu-75, and Tu-85, the last of Tupolev's piston-engined bombers. Also described are various experimental versions, including the Burlaki towed fighter programme.

Sbk, 280 x 215 mm, 128pp, 225 b/w and 9 colour photos, plus line drawings
1 85780 142 3   **£18.99/US $27.95**

Red Star Volume 8
RUSSIA'S EKRANOPLANS
Caspian Sea Monster and other WIG Craft
Sergey Komissarov

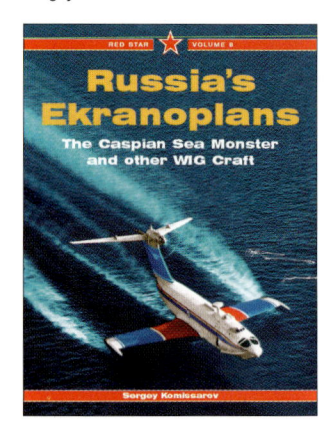

Known as wing-in-ground effect (WIGE) craft or by their Russian name of ekranoplan, these vehicles operate on the borderline between the sky and sea, offering the speed of an aircraft coupled with better operating economics and the ability to operate pretty much anywhere on the world's waterways.

WIGE vehicles by various design bureaux are covered, including the Orlyonok, the only ekranoplan to see squadron service, the Loon and the KM, or Caspian Sea Monster.

Sbk, 280 x 215 mm, 128 pages 150 b/w and colour photos, plus dwgs
1 85780 146 6   **£18.99/US $27.95**